FIRE
IN A
WIRE

"A bold thesis… weaves anecdote and analysis… provocative … futurism by way of personal odyssey."
  -- Brian K. Mahoney, Chronogram Magazine

"Congratulations Steve Nelson for your remarkable time and effort to capture this moment in human history! Well done."
  -- Rouzbeh Yassini-Fard, inventor of internet over cable

"A provocative invitation to further thinking... accessible and enjoyable... Steven Reed Nelson is a free ranging thinker."
  -- Charles Giuliano, Berkshire Fine Arts

"He isn't just 'some guy' with an idea… a long career in communications and technology… his theory is worth reading."
  -- Jennifer Huberdeau, The Berkshire Eagle

FIRE
IN A
WIRE

**Electricity Empowers
Human Evolution
Beyond Homo Sapiens**

STEVEN REED NELSON

Massaemett Media

Published by
Massaemett Media
330 Cole Avenue, Suite B406
Williamstown, MA 01267

ISBN: 978-0-578-89009-8
Library of Congress Control Number: 2025913508

Cover by Amanda Hill / Studio Two
Front cover hand-and-sparks image by ra2studio © 123rf.com
FIRE IN A WIRE cover font by www.textstudio.com

For Jan, always

CONTENTS

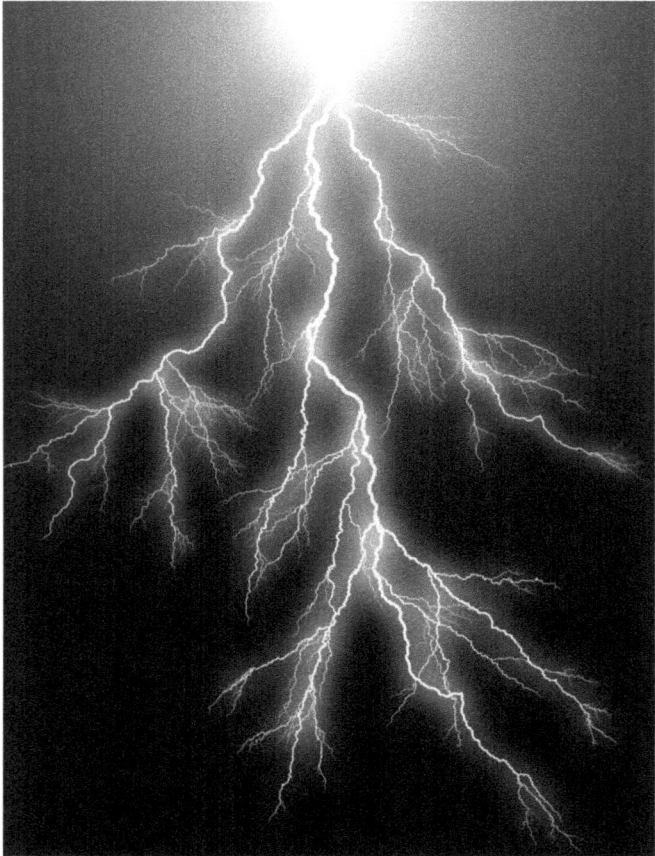

The day when we shall know exactly what electricity is will chronicle an event probably greater, more important than any other recorded in the history of the human race. The time will come when the comfort, the very existence, perhaps, of man will depend upon that wonderful agent.

-- *Nikola Tesla, inventor*

Charles Darwin

INTRODUCTION

This is a book about human evolution. Not just about how we evolved long ago, but how we are evolving now. Right now.

I was not educated to be an evolutionist. Nor was Charles Darwin. In 1831, having just graduated from college at Cambridge University, he signed up as the geologist and naturalist aboard the HMS *Beagle* on a nearly five-year voyage around the world. He wondered about the many variations among the species he observed at sea and ashore. When he returned home, he continued wondering and researching. It was not until 1859 when he published his conclusions in his landmark book *On The Origin of Species*. It caused a sensation.

Darwin observed that many more members of a species "are born than can possibly survive," so there is a "recurring struggle for existence." If, through natural variance, a being should have a trait slightly more favorable for its survival, then it will be "naturally selected," and that trait will be inherited by its progeny. He never used the term "evolution" or "evolve" until the very last word of the book, when he said:

> There is grandeur in this view of life...endless forms most
> beautiful and most wonderful have been, and are being, evolved.

To Darwin, evolution was an ongoing process including, we can only assume, for human beings. When I write in this book about humans continuing to evolve today, that is consistent with his view of life. We are being evolved.

Like Darwin, I too have for years been observing "endless forms most beautiful and most wonderful," by which I mean the myriad applications of electricity, like the computer I'm using to write this. Our lives today are vastly different than those of people who lived two hundred years ago, before we first harnessed electric power. The changes brought about by electricity are not merely a matter of progress. We are undergoing an evolutionary transformation as human beings. Our species *Homo sapiens*, which has existed for 200,000 years or more, is evolving into a dynamic new species, *Homo electric*.

In this volume I hope to shed light for a general reader on how this is happening, and why it matters. The book is part fact and part speculation, part discovery and part deduction, part popular science and part pop culture, part memoir and part manifesto. It is at times profound and at times profane, at times historical and at times futuristic. It is based on a simple yet revolutionary idea: electricity is affecting human evolution, profoundly changing who we are and who we will be.

In the course of my writing, I learned that not only are we as a species evolving, but so too is the concept of "evolution" itself. New scientific thinking is changing our understanding of how human evolution works. It supports my view that we "have been, and are being, evolved" in the Electric Age.

Natural selection as posited by Darwin is not the sole driver of human evolution today. We no longer live in a state of nature struggling to survive, like swarms of insects, schools of fish, flocks of birds or groups of mammals. We are unique beings subject to different rules and different forces, many of our own making. We are acquiring the power to control our own evolution. We must if we are to survive.

During his voyage on the *Beagle*, Darwin ventured into the Andes. Many years ago, I too journeyed there, living among descendants of the Incas, without electricity. That is where my voyage of discovery about evolution began.

4

Darwin party crossing the "Bridge of the Inca" in the Andes (illustration from Charles Darwin, The Voyage of the Beagle*)*

The author on a mountain pass at 14,000 feet in the Andes

PART ONE

Living Without Electricity

If it weren't for electricity, we'd all be
watching television by candlelight.

-- George Gobel, comedian

Peruvian Airlines DC-6

BACK TO THE PAST

My flight was delayed. Again. "Mechanical problems."

I was on a layover in Miami en route to Lima, Peru and my final destination, Vicos, a remote community at 10,000 feet in the Andes. I'd left LaGuardia at midnight and arrived in Miami at 3 a.m. on 6/19/61. My connection to Lima had been due to depart at 7 a.m., now hours ago. Shaky after the flight, my first ever, I needed to settle my stomach and my nerves, facing 12 hours on a Peruvian Airlines propeller-driven DC-6.

I'd been up all night, but was too restless to try dozing off in the hard plastic waiting area chairs, and wandered aimlessly around the spartan airport. Hoping for a breath of fresh air, I stepped outside the terminal, but a blast of Florida heat and humidity drove me back inside. Suspended in time and space, all I could think about was the journey I was on. That is, assuming the plane was able to take off and I was still willing to get on it, despite my anxiety about what lay ahead.

The East Coast of the U.S. was in the same time zone as Peru on the West Coast of South America, so I wouldn't have to reset my watch. But I was about to go back in time to a place totally removed from my middle-class American life. They didn't even have electricity.

The flight to Lima finally departed at 2 p.m., seven hours late. I was aboard, with a classmate from Cornell University. We were among six students from Cornell, Harvard and Columbia awarded grants to spend the summer in Vicos doing anthropological field research. I had only a vague idea of what it would be like to live there; we had no formal training to prepare us. You just got on a plane and showed up.

The two of us were the only North Americans on board, surrounded by Peruvians before we even arrived in Peru. The DC-6 plodded along, its propellers droning loudly. After a re-fueling stop in Panama, it got dark. Looking out the window there was no way I could tell where we were, over land or sea. For food service, they handed me a ham sandwich.

Lima was often under a dense cloud cover. When we reached its airspace, with no radar to guide us in, we circled for more than an hour. The pilot had to bide his time until he could see the lights of the city through an opening in the clouds, then dive down through it. When he finally did, we had no warning. I fumbled for the airsick bag. Shit, too late. As my body plunged toward earth, it ejected the partially di-gested remains of the ham sandwich, which landed on my lap. The Peruvians did their best to ignore me.

We touched down at 2 a.m., twenty-six hours after leaving New York. I disembarked unsteadily down stairs leading di-rectly onto the tarmac. One whiff of the fragrant tropical air and I knew I was somewhere *else*. At the gate I presented my passport and the immunization record of the multiple shots I needed to enter the country. Peru and Vietnam were the only two places in the world requiring visitors to be vaccinated for bubonic plague, the "Black Death" which killed 50 million people in medieval Europe, half the population. I was relieved to have made it to Peru. I had no expectation of ever going to Vietnam. I hardly even knew where it was.

Our student group spent two days in Lima getting to know each other and exploring the city, the capital of Peru. It was founded in 1535 by the Spanish *conquistadors* after they over-threw the Inca empire. Lima was growing rapidly due to the influx of indigenous Andeans coming down from the mountains in search of a better life. They lived in shantytowns on the outskirts of the city. We visited one on the edge of a large garbage dump. The residents foraged there for materials with which to build shacks – sheet metal, scrap wood, cardboard – and for scraps of food, competing with scavenging feral dogs. Just like back home, they had no electricity.

Our group leader was a graduate student from Cornell, Paul Doughty, who with his wife Polly had spent considerable time in Peru. He took us to see the ruins of Pachacamac, in the desert near Lima. First settled around 200 AD, it fell under the control of the Incas in the 15th century. Unlike the monumental stone structures of famed Machu Picchu, the site was built with adobe and small stones. Its buildings had eroded over the years, like crumbling sand castles. We were the only visitors. It was eerily quiet, just the sound of the wind.

Mummification had once been a highly developed practice in Pachacamac. In the extremely arid climate, mummies were well preserved. Most of them had already been removed from the site, either by archaeologists or grave robbers. Among the remains found there were those of several young women, human sacrifices with cotton garrotes still looped around their necks. The gods were appeased.

Wandering around the ruins, we came upon a shallow pit with some bone fragments and remnants of old handwoven cloth scattered about. Paul stepped in, reached down, brushed away some sand and, to our amazement, lifted out a human skull. He held it up, like Hamlet contemplating Yorick. The colorful pieces of cloth were remarkably well preserved, because it hadn't rained there for centuries.

Paul Doughty holding skull he unearthed at Pachacamac

Site of deadly sandslide on the Pan-American Highway

We left Lima for Vicos in Paul's Jeep station wagon, traveling on the Pan-American Highway. It runs nearly 20,000 miles from Prudhoe Bay, Alaska, north of the Arctic Circle, to Argentina at the southern tip of South America, hugging the Pacific coastline in Peru. Our route would take the highway north for several hours, then head east up into the mountains and across a pass before descending into the high Andean valley where Vicos was located.

Not far out of Lima we came upon another place touched by violent death. The road abutted a sandy hillside to the east, with a steep drop to the ocean to the west. A southbound bus on the ocean side of the road had been swept over the edge by a sand slide from the uphill side. All the passengers were killed, which was headline news when we were in Lima. As we neared the scene of the accident, the road narrowed to a single lane where a bulldozer was clearing it.

Having to stop and wait, we got out of the Jeep, and could see the wreckage of the bus far below, lapped by the Pacific. The bodies had been removed but it was an unsettling sight. We were well aware that Peru was prone to epic disasters: massive floods, enormous avalanches and powerful earthquakes. Here was another one to think about, as we faced a long journey up the coast with the hillside looming above us.

We breathed easier when, 200 miles north of Lima, we stopped for lunch at a Chinese restaurant. In the 19th century, Chinese emigrants came to Peru to work on building the railroads, as they did in the western United States. Best wonton soup I've ever had, with lots of seafood, and a mix of Chinese and Peruvian spices. Where we were going there wouldn't be any Chinese restaurants, or any restaurants at all.

IN THE ANDES

After lunch we turned off the highway to begin our ascent into the Andes. The road into the mountains was one switchback after another as we gained elevation. Just when we thought we were finally seeing the top of the mountain range, we'd come around a bend and see even higher mountains ahead. The road was unpaved, not really two full lanes wide, with no guard-rails. As we climbed along the outside of the road, a bus or truck coming down toward us would have to squeeze up against the hillside to get by, forcing us perilously close to the edge. Sitting by a window on the passenger side of the Jeep, I looked down at a drop-off of hundreds of feet, fearing that we too might soon make headlines.

We finally reached the pass, at about 14,000 feet, and could see towering snow-capped peaks in the distance across a deep valley. We stepped out of the Jeep to look around and take a breath of the thin cold air. The land was rock-strewn and devoid of any greenery. One of the students complained of chest pains from the altitude. We were all relieved when they subsided as we descended into the valley, the Callejon de Huaylas, toward our destination of Vicos, 4,000 feet below.

Callejon means "alley," and it is just that, a passageway from south to north between the Cordillera Negra, or Black Mountain Range, to its west and the Cordillera Blanca, White Mountain Range, to its east. The latter includes fifty glacial peaks which, at over 18,000 feet high, are taller than every mountain in North America except Denali in Alaska. It is crowned by Huascaran, at 22,205 feet the tallest mountain in the Earth's tropical zone.

The floor of the Callejon is about 13,000 feet above sea level at its southern end and about 6,500 feet at its northern end. The Santa River runs through it, arising from a high mountain lake and flowing north toward the Equator until, at the end of the Callejon, it turns west seeking the Pacific.

In the late 1940's the Peruvian government proposed to dam the river to produce hydroelectric power in the Callejon. This caught the attention of the Peruvian anthropologist and Cornell graduate student Mario Vazquez as an ideal opportunity to study the impact of rural electrification on people without power. As it turned out, the dam was not built, but the *hacienda* (plantation) of Vicos became the focus of another anthropology project, which was what brought us there.

Descendants of the Incas, the Vicosinos had for centuries, since the Spanish conquest of Peru, been subjugated under a system of feudal servitude. In exchange for living on the land and farming it, they owed the landlord three days a week of their labor, enforced with a whip by an overseer on horseback. Every five years the corporation which owned the 30,000+ acres of land offered up the lease on the Vicos hacienda for renewal, at a cost of about $600 per year. That entitled the leaseholder to the labors of the 2,000 people living there, and their working animals too.

In 1952 Cornell Professor Allan R. Holmberg and Vazquez founded the Cornell-Peru Project, which took over the lease and freed the Vicosinos from their serfdom. Nine years before the creation of the Peace Corps, the focus of the Project was on working closely with the Vicosinos to improve health care, education and farming practices, as well as their self-esteem. By 1957 the community was self-governing.

Had the hydroelectric dam been built, it would have been an event not unlike what many rural communities in the U.S. experienced with New Deal projects. In the early 1930s, only 3% of farm homes had electricity. Twenty-five years later, 90% did. Of course, we've all gone without electricity for a few hours during a power failure. Maybe you've been off the grid on a camping or fishing trip for several days. But that was just a break from your normal life. Cook over a campfire. Have a few beers. Hope your phone battery doesn't die.

When our student group arrived in Vicos and exited the Jeep in the village plaza, it was like stepping back in time. Until little more than a century ago, our ancestors had lived for millions of years without electricity. Since they began to string telegraph wires in the mid-19th century, the uses of electricity have proliferated to an astonishing extent. We were leaving all that behind. I wasn't thinking about it at the time, just trying to get by day-to-day in this strange yet beautiful land, but living in Vicos without electricity eventually helped shape my views about the effect on human evolution after we learned to control electricity. In Peru, I lived in the *before*.

After the long flight from New York to Lima, and the long drive from Lima to Vicos, I was about to depart on the final and most eventful leg of my trip, walking from the plaza to the home of the Vicosino family where I would be staying. It was only about 1/3 of a mile away, but the biggest leap of all, beyond western civilization. My life was now centered around the home of Nestor Sanchez Bautista, a respected village elder in his seventies, and Carmela Tafur Copitan, sixty something.

They had four sons, the youngest of whom was born when Carmela was in her fifties. People who live in high mountains,

Nestor brewing corn-based beer called "chicha"

Carmela (center) wearing her finest at her son's wedding

like Andeans and Tibetans, evolved to have higher levels of hemoglobin in their blood and increased lung capacity, for breathing air with less oxygen. Fertility rates are lower at such altitudes, but natural selection compensated by extending the age when they are able to reproduce.

I lived at Nestor and Carmela's in a small adobe room with a cot, kerosene lamp and dirt floor, clustered with a few other small rooms around a little courtyard. The eat-in kitchen, another small room, had an open cook fire and a hole above it to let out smoke. This was nothing like the cookouts over charcoal fires we had on Long Island. Vicosinos rarely ate meat, occasionally roasting guinea pigs they raised or, for special community occasions, goat meat cooked on hot rocks, like a clambake. A native grain *kinuwa*, unknown at the time outside Peru, was a regular source of protein. The mainstay of their diet was potatoes, usually prepared in a watery soup. They came in many colors: white, yellow, red, blue, purple. Some were eaten raw, crunchy and sweet.

The potato was first cultivated in southern Peru and northern Bolivia about 10,000 years ago, when people there first began living in agrarian settlements rather than as hunter-gatherers. Potatoes were key to their evolution as farmers. The Vicosinos grew them in nearby fields, plowed with oxen, but potatoes could grow at even higher altitudes. The *puna* is a high Andean grassland which runs from Peru to Argentina, above the tree line and below the glaciers, at altitudes of 12,000 to 16,000 feet. Soil nutrients and adequate moisture make it highly suitable for growing potatoes.

One of the other students and I went with Nestor to the puna for a couple of days, a daunting five-hour horseback ride over steep rocky terrain. We experienced how Andeans lived thousands of years ago, sheltering at night under a huge rock overhang with their sheep, pigs and dogs. A fire warded off any pumas with designs on the livestock. In the morning we set off to dig for potatoes growing helter-skelter among clumps of wild grasses, as they had when long-ago Peruvians foraged for them. The Vicosino men took occasional work breaks to chew coca leaves, which grew wild in the Callejon.

Had my mother known about my diet in Peru, she would have been horrified. She couldn't do anything about that, but she did send me food for thought every Monday: the Week in Review and Sports sections from the Sunday *New York Times*. There was no mail delivery in Vicos, so Paul drove to the post office in the nearest town to pick it up. I was surprised to get the first package from her, which I wasn't expecting, and amazed that it arrived just three days after she mailed it.

Reading the *Times* every week, I followed the escalating confrontation between the western Allies and the Soviet Union over control of Berlin. The Soviets were building a wall around West Berlin, and it seemed like there was an increasing chance of nuclear war. Thinking about that possibility, I feared there would be nothing left to go home to.

I knew that I could survive in Vicos if I had to. Over the summer I had acclimated to the altitude, adapted to the Vicosino diet and way of life, and bonded with the Sanchez family. They even gave me a suit of clothes made for their oldest son, who now lived in Lima and wore western dress. The homespun wool was itchy, but it kept you warm on cold nights in the mountains Still, with no electricity, the Andes was not my natural habitat. I had not evolved to live there.

Foraging for potatoes growing wild on the puna

The author (left) with a Vicosino

At the end of the summer, I moved into the Cornell head-quarters building on the village plaza to prepare for departure at dawn. In the middle of the night I was awakened by the furious ringing of the bell in the chapel. Throwing on my clothes and stumbling outside, I saw a crowd had gathered and was looking up at the cause for alarm: the full moon was turning deep red in a near-total eclipse. With no sources of ambient light, the lack of moonlight left us in deep darkness. People have long taken events like eclipses as omens. For me it marked my passage back to the world I had come from, where we lit the night with electric light.

When I got home to Long Island, my mother made me a big steak. I devoured the meat, but please, hold the potatoes. Over the next few days I devoured media: radio and TV. Then I returned to Cornell, and the Berlin Crisis was soon defused.

I felt distant from college life; my thoughts often wandered beyond the confines of the classroom to the vast spaces of Peru. "A mind that is stretched by a new experience can never go back to its old dimensions," wrote Supreme Court Justice Oliver Wendell Holmes, Jr. Mine had been stretched and pushed and twisted in the Andes, living without electricity.

It would be many years until I fully grasped the signifi-cance of what I'd experienced there, living as people had for millennia since hunter-gatherers took up agriculture. By then, 2019, I'd become deeply involved with something made pos-sible by electricity and, as I would realize, essential to the ongoing evolution of humans: the internet. Once again I found myself living without a modern necessity, a good internet connection, but this time I was actively doing something to change that.

PART TWO

Connections

Every single thing in the world that was made by anyone started with an idea. So to catch one that is powerful enough to fall in love with, it is one of the most beautiful experiences. It's like being jolted with electricity and knowledge at the same time

-- David Lynch, filmmaker

Washington at Valley Forge

GETTING CONNECTED

As I gazed out my window, with the sun setting behind the Berkshire hills, a lightbulb was about to flash on. In my head.

I was thinking about the project I'd been working on for nine years: to build fiber-optic networks to bring the internet to several small western Massachusetts towns, including mine, Washington, population 494. It was founded in 1777, the year George Washington led his army to their winter encampment at Valley Forge, where they lacked basic necessities like food, clothing and equipment. My neighbors and I lived in our own kind of Valley Forge, lacking a basic necessity of modern life: reliable high-speed internet. It was the winter of our disconnect. Too many winters, too many summers.

These sparsely-populated towns were not attractive markets for the large companies providing internet service, so we had to get by with inferior technology. DSL (digital subscriber line) ran over the "twisted pair" wires which carried telephone calls. You could get a downlink of maybe five or ten megabits a second, and an uplink of perhaps three megs, not bad ten years ago. The problem was that a DSL system could only serve a small number of customers, and you had to live within about 3½ miles of the building which housed the equipment to provide service. The further you were from that so-called central office, the weaker and slower the signal got. Beyond the limit, you were shit out of luck.

In that case, you'd need a satellite dish to get internet. The service was worse than DSL, maybe two or three megs down, less than one measly meg up. It worked by connecting to a satellite 22,236 miles above earth in geosynchronous orbit, at a fixed position relative to the earth below. When you clicked on a link to go to a website, say Google, your dish sent a signal up to the bird, which then relayed it down to another dish on a wired terrestrial network, which retrieved the link to Google and sent it back up to the satellite and then down to your dish. That's how Google showed up on your screen.

Even though the signals were transmitted at the speed of light, 186,000 miles per second, you're looking at 22,236 miles up and down to send your click to the internet, and then 22,236 miles back up and down to deliver Google. That delay, called latency, while a fraction of a second, caused all sorts of problems. Plus, the signals going back and forth through the atmosphere were subject to interference by weather systems. I had so many dishes on my roof over the years that my house looked like a CIA outpost. I could often tell a storm front was approaching when the internet got hinky. In the middle of a blizzard, if you wanted to find out how much snow to expect and when it would end, sorry, no internet connection. Try again later, after the snow melted off the dish.

To redress the problem of poor internet service, in 2010 a group of small towns in western Mass. organized a municipal cooperative called WiredWest. Its mission was to plan and advocate for fiber-based internet, and then to provide it when the glorious day arrived at long last. A group of eight or so volunteers was leading the effort. I was chosen to chair the Legal/Governance Committee of the coop Board of Directors.

None of us knew much about building and operating fiber networks, but working together we learned. After going down many dead ends for financing, by 2016 we had settled on a plan that called for the towns to put up about two-thirds of the funds needed to deploy fiber, with the state providing the other third. At town meeting after town meeting, people voted overwhelmingly, in some cases unanimously, to raise their taxes to pay their share. They wanted high-speed internet badly. With fiber they would get symmetrical service, the same speed down and up, of 1000 megabits per second, a gigabit.

Continuing to gaze out my window that afternoon, I recalled when construction of fiber networks finally got underway in 2018. In each town a small prefab concrete building was installed to house the equipment for operating the town's network. A flatbed trailer brought the "hut" to its site next to the Washington town hall, and a giant crane lifted it into place. That drab little building was the castle of our dreams.

Fast and reliable internet would significantly improve the lives of everyone in these small towns. Teachers sometimes assigned kids videos to watch online as part of their homework, but until now they couldn't. Parents couldn't take continuing education courses for their jobs, or book vacations. You couldn't watch movies and TV shows. You couldn't play multiplayer online games. Shopping online was difficult, banking was out of the question. You couldn't do a video call with family or friends out of town. But all that changed, with a hairlike strand of glass which connected us to the world, just when the COVID pandemic physically isolated us.

It was not all that long ago when a gigabit connection, a billion bits per second, would have seemed like science fiction. Throughout the 1960s and well into the 1970s, the standard speed for modems was 300 bits per second. By 1976 it had increased to 1200 bps, then to 2400 by 1984. Ten years later, it hit 28,800 bps, commonly called 28k.

The equipment hut being lowered into place in Washington

Satellite dishes removed from my roof after getting fiber

By then I was producing business-to-business videos about the "hot trends and cool products" in cable TV and consumer electronics, and began streaming them so people could watch on their computers. The quality of video streamed at 28k was awful, but you could watch a video without first having to download it, which could take hours. I knew that inevitably video and internet technologies would evolve. They sure did.

The widespread availability of high-speed internet service in the U.S. was led by the cable TV industry, but in a sense that was a fluke. For decades, no one in the cable business was thinking about using their systems for internet access. People went online through dial-up service over phone lines, accompanied by the weird sounds of modems "handshaking."

Actually, one guy was thinking about it, Rouzbeh Yassini-Fard, an engineer born in Iran who emigrated to the U.S. in 1977 to attend West Virginia University. In the late 1980s he had a crazy idea: to provide internet service over the same coaxial cables that carried video. Can't be done, people said.

But Yassini-Fard figured out how cable networks could handle the two-way traffic required for internet access. To do it he invented the cable modem, and founded a company called LANcity to market it. Then he led the effort to develop the technical standards necessary to make high-speed internet service possible on any cable system with modems made by any manufacturer. Goodbye 28k. Hello broadband.

Today, viewing of streaming video has surpassed total combined viewing of broadcast, cable and satellite television. There are now about a billion videos that you can watch on YouTube. By comparison, the Library of Congress has 25 million books in its collection, the largest in the world.

MAKING A CONNECTION

Still gazing out the window, I thought about how electricity has changed the way we live. I recalled as a little kid the first time I saw TV, in a radio store in the Bronx in 1947, the first telecast of the World Series, the Yankees and the Dodgers.

New York radio ad sales rep Ralph Baruch remembered his first time. Walking on Broadway one evening, he noticed a crowd of people watching a wrestling match on a TV set in a store window. When he came back later, some people were still watching the TV. Curious what was on, he went over to see. The match was over, but he was amazed by what had them transfixed: the motionless test pattern that TV stations screened when their regular programming ended late at night. Hmm, he thought, if they'll watch that, I've got to get into this business. He later became a prominent cable TV executive.

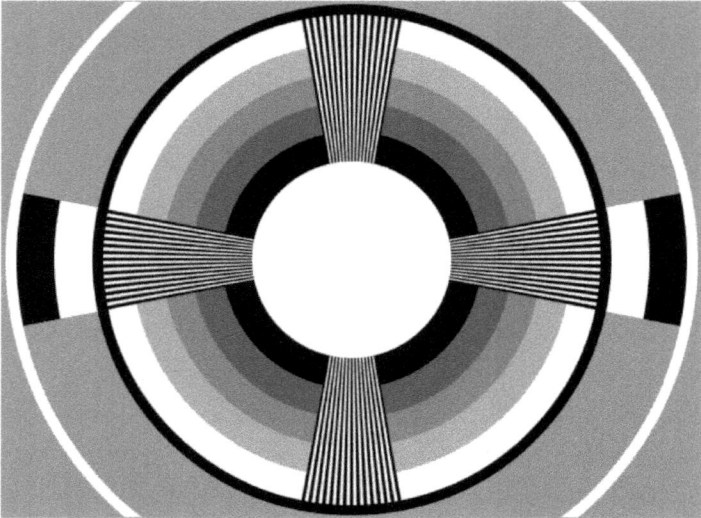

As the sun set, I recalled my childhood on Long Island in the 1950s. We had all the "modern" electrical conveniences: TV, radio, record player, refrigerator, stove, a few appliances. Not all that much, really, compared with our homes today. But it wasn't that long ago when we had none of them.

The word "electricity" did not even exist until 1600, when the English scientist William Gilbert coined the term "electricus" from the Greek word *elektron* for amber. The ancient Greeks knew how to produce static electricity by rubbing an amber rod with a silk cloth. The words "electric" and "electricity" first appeared in print in 1646.

It would be another two hundred years before a useful application of electricity was introduced: the telegraph. When people first got electric power in their homes in the late 19th century, the only thing they could use it for was lighting, thanks to Thomas Edison's incandescent invention. But the simple act of turning on a lightbulb was a revelation. A lightbulb is still symbolic of a new idea, a flash of inspiration.

It was getting dark when I had a "lightbulb moment." I knew that controlling fire was a pivotal development in the evolution of archaic humans. Having lived without electricity, I made the connection between the importance of fire and now electricity to human life. I realized that ***controlling electricity is a pivotal development in human evolution***. Electricity is the new fire. Fire in a wire.

This idea of electricity enabling our evolution got me excited, so I Googled it, expecting to find lots to read. But there was nothing. WTF?!? I was on my own to figure it out. To understand how we're changing, I had to learn more about how we got here. So I went back to the beginning. Way back.

PART THREE

Bones and Stones

The universe exists for us only
insofar as it exists in our brains.
The brain is our three-pound universe...
and it runs on electricity.

-- Judith Hooper, journalist

THE DAWN OF HUMANS

First light appears in the early morning sky and spreads over the savanna of East Africa. The busy chirpings of birds in the scattered trees pierce the cool air. In the distance a large predator roars, signaling its family that another day of hunting for prey is about to begin. The sound causes an uneasy shuffling of hooves among a herd of grazing wildebeest.

Under a large rock overhang, a band of humans camped there is also stirring. An old woman is first to move about, adding some dry grass and twigs to coax flames from embers banked overnight. The warmth from the fire encourages the others to arise. They are dressed in animal skins and are barefoot. Their language is simple, just a few words and phrases, accompanied by hand and facial gestures. There are only a few bones left from a recent kill, so to fill their bellies the men, two women and the teenage boys will head out to hunt game. The rest of the women, an old man and the children will forage for plants, nuts, berries, roots and insects.

100,000 years later, I'm sitting at my computer, foraging in my mind for thoughts and words to put on this page. I'm a hunter-gatherer of information and ideas. I can easily imagine those ancient Africans in their shelter, having spent those two nights with Vicosinos under a rock overhang in the puna, rising at first light as women stoked the fire to cook breakfast before we set out to forage for potatoes.

But it is more than just personal experience that enables me to understand who those ancient people were and what their lives may have been like. I have the benefit of the knowledge we've gained since the scientific revolution began some five hundred years ago. It has empowered us to understand life as it was, as it is, and as it perhaps will be.

On that long-ago African morning, those archaic humans could not have known of the time some 100,000 years or more before them when their species *Homo sapiens* emerged. Nor could they have known of those creatures which preceded them over many tens of millions of years, small mammals leading to primates leading to apes leading to the hominins who eventually became *Homo sapiens*.

Our ancestors came down from the trees to find their footing on two legs, then learned how to work together to chase a large animal until it was overcome by fatigue. They picked up stones to break open animal bones and nuts, then learned how to fashion stones into tools for cutting and chopping and scraping. They tended the fires left from lightning strikes, then learned how to make fire themselves.

For that band of humans in the camp, it was a day like all other days, moving about the African plains in search of the food they needed to survive, finding a place each night where they could make their fire and be safe from predators. And so it went for tens of thousands of years. Yet by about 60,000 or so years ago, they began to move in new directions, to the north out of Africa, and around the world.

It's not clear why they migrated out of their comfort zone, if we can call it that. To follow game? Driven by a changing climate? Curiosity about what lay over the horizon? Whatever the reason, they did migrate, following their destiny to eventually become modern humans. To understand who we are now and who we are becoming, we need to understand who we once were and how we got here.

THE DESCENT OF MAN

In *On the Origin of Species*, Charles Darwin never said explicitly that humans were descended from apes. Perhaps he was concerned that such a sensational notion would get all the attention and that the details of his theory, which was controversial enough, would get lost in the fuss. Still, the implication was there for anyone who thought about it, because in his view all forms of life were descended from a common ancestor, including humans.

In his second landmark work on evolution, *The Descent of Man, and Selection in Relation to Sex*, published in 1871, Darwin came out and said it: humans are descended from apes. By then, it was not so shocking a concept because people had been talking about *Origin* and its implications for twelve years. Yet there were those who could not accept the idea and howled in protest, like the monkeys they had once been. Some still cannot accept it, believing that the world was created 6,000 years ago as told in the Book of Genesis. It gets tricky trying to herd all the dinosaurs into that origin story.

When Darwin wrote *Origin*, the descent of man from apes was still a theory without proof. But just three years before, and unknown to Darwin at the time, some workmen digging in a valley near Dusseldorf, Germany uncovered a large number of what seemed to be oddly shaped human bones, including part of a skull. A local schoolteacher realized that the bones were not quite like those of modern humans and brought them to an anthropologist. After reading *Origin*, they agreed that the bones were those of a different kind of human. Their conclusion met with some initial resistance, doubters arguing that the bones were merely deformed. But it was not long before it was widely accepted that they did in fact prove, for the first time, the former existence of another species of human. Many more bones were found, and they were classified scientifically as *Homo neanderthalensis*, the man from Neanderthal (Neander Valley). The valley was named for the 17th century German theologian, Joachim Neander, whose last name can be translated as "new man."

Since I'll be throwing around a lot of *Homo* this and *Homo* that in the next several pages, before I do let me say a few words about "taxonomy," a method of classification. In biology it is used by scientists to name and describe all living things, past and present, and to depict the relationships among them, using a hierarchy of Latin names. Its invention is attributed to the Swedish biologist Carl Linnaeus, although he did not devise the concept of classification (the ancient Greeks did) and his system has since been altered.

Basically, Linnaeus divided organisms into two kingdoms, animals and plants. Within the animal kingdom, he assigned them to six classes: *mammalia* (mammals), *aves* (birds), *amphibia* (amphibians), *pisces* (fishes), *insecta* (arthropods) and *vermes* (worms). He then subdivided the classes, and subdivided the subdividees, and so on.

Linnaeus published the fundamentals of his taxonomy in 1735 in the first edition of *Systema Naturae*, a mere twelve pages long. By the 10th edition in 1758, it classified 4,400 animal species and 7,700 plant species. People from around the world kept sending him information about their local flora and fauna. To keep track of it all, he invented the index card. Today you'd need 2.3 million cards to list the known species.

As you can see, Linnaeus was big on Latin, and through all the changes to his system since then, one thing has remained unchanged: his use of two Latin words to name each specific species. For example, the familiar name for human beings is *Homo sapiens*, classified within the genus *Homo* ("man" or "human" in Latin) as the species *sapiens* ("wise" or "intelligent"). Just to complicate matters, sometimes we are referred to as the subspecies *Homo sapiens sapiens* in order to distinguish modern people from earlier versions of our species. I'm not sure that adding another "sapiens" means we're any wiser.

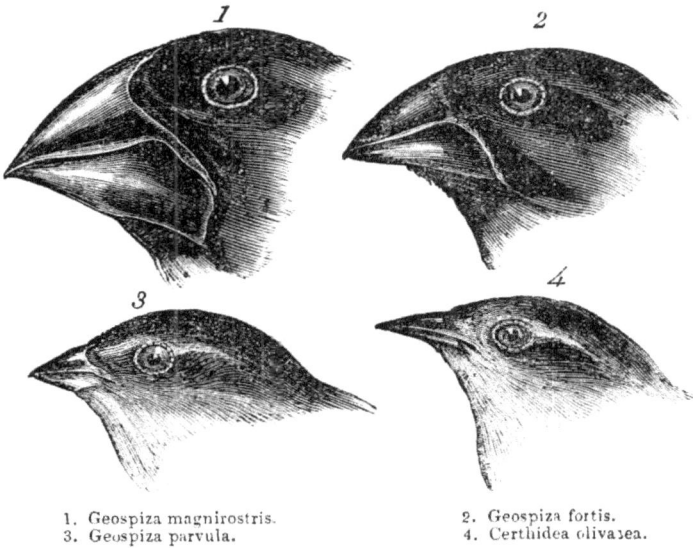

1. Geospiza magnirostris.   2. Geospiza fortis.
3. Geospiza parvula.   4. Certhidea olivacea.

Darwin's classification of several species of finches,
illustrated in The Voyage of the Beagle

Once *Origin of Species* was published, the Linnaean tax-onomy turned out not to be at all useful for depicting the evolutionary relationships between organisms. Darwin himself used a diagrammatic "tree of life" to depict how species branched out from their predecessors. It was a concept he first used in sketches he made soon after he returned from his voyage on the *Beagle*. Later it became the only illustration he included in *Origin*, so he obviously attached great value to it.

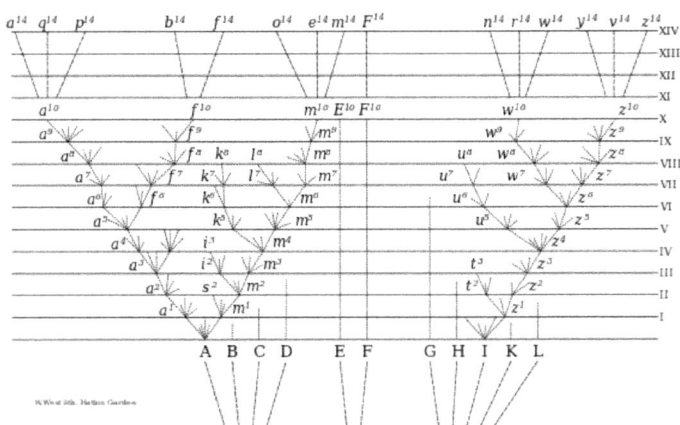

Darwin's Tree of Life diagram in On the Origin of Species

The concept has since become more complicated than we need to concern ourselves with here, but it is fundamentally how we still depict the course of evolution. So given the Linnaean nomenclature for species and a branching evolutionary tree, let's venture back into the primeval world from which we arose, and track humans as they evolved over millions of years. They left behind a trail of bones and stones, which many paleontologists and archaeologists have worked painstakingly to uncover, brushing away the sands of time.

A PRIMER ON PRIMATES

The animals we call mammals have been around for over 200 million years. For most of that time they were tiny nocturnal creatures, no bigger than squirrels. They ate insects and stayed out of the way of the dinosaurs which ruled the earth. When the dinosaurs suddenly went extinct about 66 million years ago, the populations and varieties of mammals exploded.

Primates are a group of monkeylike mammals. There is fossil evidence of primates dating back about 55 million years ago, but they may have first arisen even earlier, during the very late stages of the time of those giant dinosaurs. The evolutionary line of the great apes – gorillas, orangutans, chimpanzees, bonobos and ultimately humans – branched off from their primate relatives about 15-20 million years ago.

Looking at the line which led to humans, orangutans split off about 13 million years ago, gorillas about 10 million years ago, chimps and bonobos a few million years later. These are very approximate dates, given a limited fossil record of creatures that often lived in humid tropical environments not conducive to preservation of their remains. But we can say that the last common ancestor of humans and chimps lived about 7 million years ago, give or take a million years or so.

Chimpanzees are the closest living relatives of humans, but it's curious what became of them and us since we went our separate ways evolutionarily. They continued to evolve from that common ancestor, but they didn't evolve nearly as radically as their human cousins. That's one of the mysteries of evolution, how we happened to become who we are, a very different kind of primate, unlike any other.

HOMININS

Descendants of that last common ancestor, on the human side of the divide from chimps, are called hominins, including members of the genus *Homo* and of previous genera. Their early history is not always clear, given the limited number of fossils and their great age. I'm not going to go too deep into some of the debates about who belongs where in the hominin taxonomy, but rather introduce some key players in the drama in many acts of our evolution.

A candidate for the earliest hominin descended from that last common ancestor (as far as the fossil record has revealed to date) is a creature from Chad in central Africa called *Sahelanthropus tchadensis*, nicknamed "Toumai" ("hope of

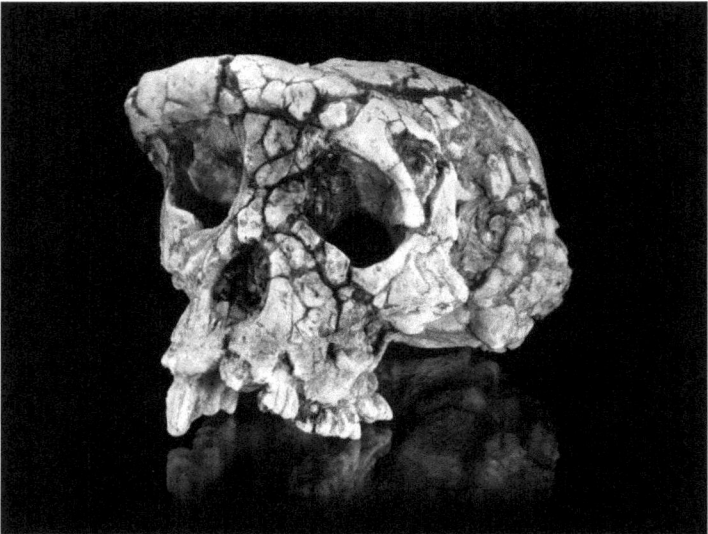

Sahelanthropus tchadensis

life"). It exhibits archaic traits such as its flat skull with a small brain case, about the size of a chimpanzee's and less than a third that of a modern human. It also has pronounced brow ridges, a forward-protruding face and large teeth, although its canines are much smaller than those of a chimp.

The main issue with classifying Toumai as a true hominin is whether its leg bones suggested "obligatory bipedalism," that is, walking on two legs all the time, a defining hominin trait. Bipedalism was a major evolutionary break from the past, when primates scrambled around in trees with grasping feet and used their knuckles to brace themselves as they moved on the ground. There is fossil evidence that Toumai was not full-time bipedal, solidifying its place as very near on the evolutionary tree to that common ancestor.

Ardipithecus ramidus also may or may not have been fully bipedal. The first fossil of this species, discovered in Ethiopia, dated to about 4.4 million years ago, but a few older bits of bone found later dated to 5.6 million years ago and was classified as a separate species, *Ardipithecus kadabba*. A nearly complete skeleton of *A. ramidus* was subsequently discovered and called "Ardi" (paleontologists seem fond of such nicknames). A female, she would have weighed about 110 pounds when she was alive 3.9 to 4.5 million years ago. Her brain was smaller than that of a modern bonobo or female chimpanzee, and only 20-25% the size of a modern human brain.

"Sexual dimorphism" is a term to describe the differing size of traits in males versus females. In chimpanzees there is marked dimorphism in the canine teeth, much larger in the male. This difference is much less pronounced in Ardi, making her more like bonobos and humans. Like Toumai, Ardi seems

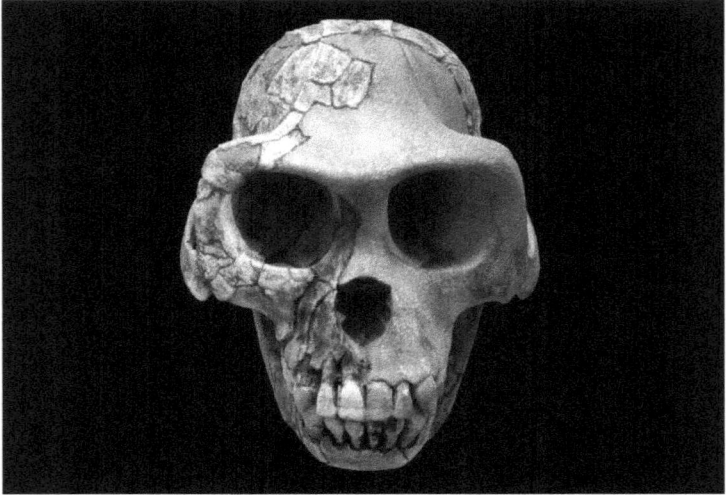

Ardipithecus ramidus

to have adapted to moving around in trees with a grasping big toe, as well as on the ground, thus calling into question her place on the evolutionary tree as a hominin.

This brings us to a genus that is truly bipedal and an important ancestor of humans: *Australopithecus*. The first specimen was discovered one hundred years ago in South Africa, the skull of a young child. The species was named *Australopithecus africanus*, and subsequent discoveries have dated it to between 3.3 and 2.1 million years ago. There have been many finds of Australopithecenes, especially in Kenya and Ethiopia, classified in six or perhaps as many as eight separate species, the oldest dating back more than four million years.

A skeleton about 40% complete and discovered fifty years ago gained worldwide fame as our ancestor known as "Lucy." She may not have been quite what The Beatles had in mind in

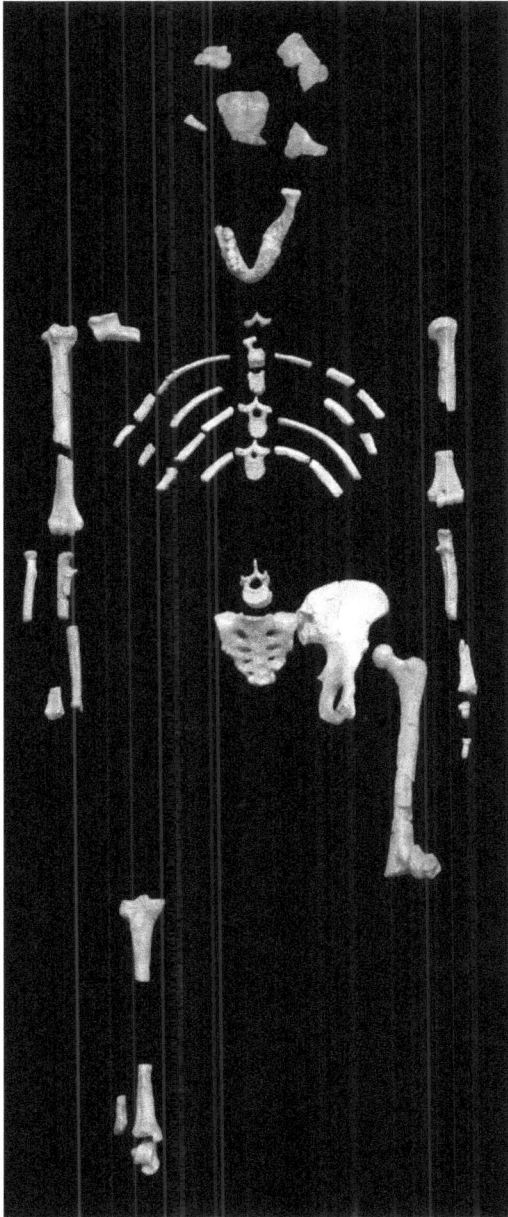

The skeleton known as "Lucy"

Reconstruction of Lucy

their song "Lucy in the Sky with Diamonds," but it inspired the paleoanthropologist who named her. Dated to about 3.2 million years ago, Lucy would have stood perhaps 3 feet 7 inches and weighed about 64 pounds. There could be no doubt from her skeleton that she walked upright. Her relatively small skull was evidence that bipedalism preceded the marked increase in brain size seen in successive stages of evolution.

A trail of early human footprints, crossing 75 feet through what was originally soft volcanic ash, was found in Laetoli in

Tanzania in 1976 by paleoanthropologist Mary Leakey. She, her husband Louis and son Richard are noted for their many discoveries in east Africa of human fossils and stone tools. The Laetoli footprints, 3.7 million years old, were probably left by three Australopithecenes, one of them a child. They provide the earliest hard proof of bipedalism, other than what could be inferred from the structure of skeletal remains.

Australopithecus was a highly successful genus, existing for over two million years before going extinct 1.9 million years ago. Or maybe not. It is debated whether the genus *Paranthropus* was descended from *Australopithecus* or should be classified within that genus. It included three distinct species who lived from 2.9 to 1.2 million years ago. Bipeds, they were characterized by their robust skulls, with a brain case at the high end of the range of Australopithecine skulls. However they are classified, they were an evolutionary dead end.

Paranthropus boisei

OUR GENUS *HOMO*

The earliest true human, *Homo habilis* ("handy man"), lived from about 2.3 to 1.65 million years ago. This overlaps with *Paranthropus* and *Australopithecus*, and some argue that *habilis* belongs to the latter genus. What makes him human was being the first hominin to make and use tools on a widespread basis as part of the culture of the species, an evolutionary milestone. Some Australopithecenes likely used tools -- simple stone tools dating to 3.3 million years ago were discovered in Kenya. For that matter, chimpanzees strip the leaves from sticks to make tools for extracting termites from their mounds, famously first observed by Jane Goodall.

Homo habilis

It was with *H. habilis* that stone tool making and use first flourished, a style of tools known as Oldowan, discovered in Olduvai Gorge in Tanzania by Louis Leakey. They used one rock to knock a few flakes off another rock, a process called knapping, to form simple choppers, scrapers and the like. They seem to have had an understanding of how certain rocks would flake when struck, but not a clear visualization of a desired end product. They just banged two rocks together and hopefully got a usable implement, perhaps only a sharp flake.

Their brains were notably larger than those of Australopithecenes, with an expanded cerebrum. There is some evidence of handedness in *habilis*, favoring the use of one hand over the other, and that is considered an indication of brain reorganization. But the arrangement of their teeth was V-shaped rather than U-shaped as in later *Homo* species, and they still exhibited prognathism, a jutting mouth.

Homo erectus ("upright man") is an important figure in human evolution, a highly successful member of our genus. It first appeared about 2 million years ago, and did not go extinct until a little over 100,000 years ago, making it the only human species to have lived simultaneously with both *habilis* and *sapiens*, the oldest and newest members of our genus (before electricity). Along the evolutionary pathway, several species came and went without being linked in a direct line to *sapiens*, but *erectus* was. Usually thought to have originated in Africa and migrated to Eurasia, early fossil finds in China known as "Peking Man" suggest that *erectus* might have originated in Asia and spread to Europe and Africa. In any case, it was the first human to appear outside Africa, and its range was greater than any other hominin prior to *sapiens*.

Homo erectus female

With a wide geographical and chronological distribution, it is not surprising that *erectus* varied in physical size and cranial capacity. It could be as big as a modern human with a brain that, at its extreme, nearly equaled ours in size. *Erectus* is known for two major evolutionary achievements. It was the first creature to control fire. This provided protection from predators; heat, critical in northern climates during ice ages; and light. Fire allowed for socializing at night, which may have contributed to the eventual development of language, although with *erectus* it was very rudimentary at best. Most importantly, fire enabled the cooking of food, especially meat, which was more easily digested than raw food and supplied the increased energy required to support development of larger brains. See Richard Wrangham, *Catching Fire: How Cooking Made Us Human* (Basic Books, 2010). When it was named in 1893, *erectus* was incorrectly thought to have been the first hominin to walk upright. For learning to control fire, it might have been better named *Homo ignis*, "fire man."

Erectus also invented a new style of tool, the Acheulean, more refined than its Oldowan predecessors, carefully worked

Comparison of Acheulean and Oldowan stone tools

on both sides with more precise edges. They were made with intentionality for the design of the desired end products, using other materials like bone to help form them. They used their stoneworking skills to create simple decorative objects, evidencing symbolic thinking and more advanced cognition.

Homo erectus gave rise to two other species of note: *Homo antecessor* and *Homo heidelbergensis*. The former, "pioneer man," appeared in Spain from 1.2 to 0.8 million years ago, and may have spread elsewhere in Europe. Once thought to have been the last common ancestor of Neanderthals and *sapiens*, it is now considered an offshoot from the human line.

That ancestral title now likely goes to *heidelbergensis*, remains of which were first found near Heidelberg, Germany, but which also existed in Africa. Fossil and DNA evidence places the Neanderthal split from the *sapiens* line as perhaps 500,000 or as much as 650,000 years ago. *Heidelbergensis* is noted for having certain physical features used in speech and hearing, advanced from *erectus* and similar to those of modern humans. This suggests they were capable of, but did not necessarily use, language.

Homo heidelbergensis

Erectus likely led to another species not important in the human line but fascinating nonetheless. *Homo floresiensis* was a species found on the island of Flores in Indonesia, sometimes called "hobbits" because they were only about 3½ feet tall. Their size is attributed to "island dwarfism." After *erectus* made it to the island but was unable to leave, natural selection favored those best able to survive on less food in an environment with limited resources. Their skulls and brains were proportionately small. *Floresiensis* arose about 190,000 years ago, so that they were alive concurrently with *Homo sapiens*, until going extinct about 50,000 years ago.

Homo floresiensis

HOMO NEANDERTHALENSIS

This brings us to the best known and most intensely studied member of the genus *Homo* (other than ourselves, of course), our closest evolutionary cousins, the Neanderthals. They lived from more than 400,000 years ago until 40,000 years ago, overlapping most of the era of *sapiens*. They had brains larger than those of *sapiens*, in a skull that was somewhat flatter and rather elongated, with pronounced brow ridges. They were slightly shorter than *sapiens*, heavy boned with a stocky body.

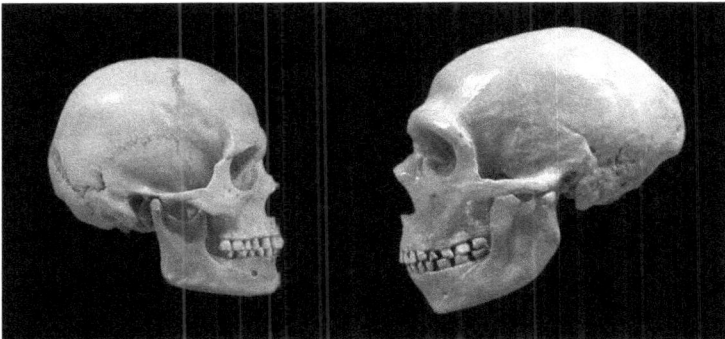

Skulls of modern human and Neanderthal

At their peak they were found from the Atlantic coast to Siberia, although their total population was quite small. Their effective population – those who could bear or sire children – varied in estimates from as few as 3,000 to perhaps 25,000. They lived in small groups, maybe as few as 10-20 individuals. Limited contact with other Neanderthal groups led to inbreeding and genetic defects, which may have caused their extinction.

Homo neanderthalensis

Neanderthal woman

For many years they were unfairly depicted as dumb brutes, the archetypical cavemen. And indeed, they often dwelled in caves and wore animal skins. But that characterization has been discredited to give them their due as more advanced in many ways. They developed more refined stone tools, known as Mousterian, the same as those used by some *sapiens* at the time. They were able to make fire, and to cook food in various ways; to create simple clothing like ponchos; to use plants as folk remedies; to care for the ill and injured; and possibly to bury their dead. They used bird feathers and talons for decorative purposes, but their art was largely limited to some scratches on walls and shells. They had the physical attributes necessary for hearing and speech, but there is no way to know if they had anything but a simple form of language, if that.

A flute made from a cave bear femur was discovered in Slovenia and has been dated to 50,000 to 60,000 years old. It has been at the center of an ongoing dispute as to whether it is a flute at all. But it does appear to be, and some scientists attribute it to Neanderthal origin, while others disagree that it was they who made it.

Whatever the case, the reaction against the "dumb brute" image has, in my opinion, led at times to a tendency to over-romanticize the Neanderthals as civilized beings. They were certainly less so than their *sapiens* contemporaries who, for example, created awe-inspiring representational art on cave walls, not just some scratches. The Neanderthals existed for nearly 400,000 years, and while they evolved beyond their *Heidelbergensis* progenitors, the last common ancestor of *sapiens* and Neanderthals, they only went so far. Look where we have come in half that time.

An example of the updating of the Neanderthal image is this statue in front of the Neanderthal Museum in Mettmann, Germany, near where those first remains were found in 1856. It depicts a Neanderthal man in a business suit. He has had a haircut and shave, and looks a bit svelte for a Neanderthal, with the waist of his suit nipped in. Was he on a Paleo diet?

The idea, I suppose, is to normalize Neanderthals. He looks just like us, sort of. But if I were a young woman in a bar and saw this guy looking my way and lurching toward me, I'd head for the ladies room pronto.

The biggest unanswered question about Neanderthals is why they went extinct about 40,000 years ago. Many theories have been advanced to explain their disappearance. Perhaps fitting for our worldview today, climate change is one theory, in this case a colder climate. But Neanderthals had been living in more northerly climes for hundreds of thousands of years, and their large bulbous noses were likely an adaptation for breathing cold, dry air. Who would you bet on to survive the cold: Neanderthals? Or *sapiens* recently arrived from Africa? It turned out to be bad news for the Neanderthals that *sapiens* were able to adapt successfully to the colder climate.

We can't ignore the woolly mammoth in the room: wherever *sapiens* appeared, Neanderthals disappeared. One explanation is that the newcomers brought with them diseases for which Neanderthals had no resistance, like the indigenous populations in the New World who were decimated by diseases brought there by the first European settlers. But this is a phenomenon that would occur over several years, perhaps decades, not over millennia, as was the case for Neanderthals.

To understand the impact on Neanderthals of *sapiens* spreading throughout the world, we have to deduce what sort of beings those early *sapiens* were. Just as we look back to our archaic forebears to glean insights into humans today, we can look at modern humans for clues about what those forebears may have been like.

Consider the past several thousand years of recorded history. We have documented accounts of human behavior showing *sapiens* to be capable of the most violent and murderous behavior of any species ever to have existed (with the exception perhaps of some species of ants). Hitler, Stalin and

Mao killed tens of millions, with the devoted support of their countrymen, and there are many more instances, if not on so massive a scale, of barbarous behavior throughout our past.

Humans are aggressively territorial, as we see today, for example, in Putin invading Ukraine and unleashing massive missile and drone attacks on its population; Netanyahu grasping territory in the West Bank and attacking Gaza, choking off food supplies there; and Trump setting loose ICE agents to seize and deport immigrants, with little concern for their legal status as residents. These are but a few examples from the heinous history of *Homo sapiens*, which we can't ignore when we ponder the fate of the Neanderthals.

Another theory argues that Neanderthals may have disappeared as a species by becoming assimilated into *sapiens*. Are we supposed to imagine a band of Neanderthals approaching their *sapiens* counterparts and, with their limited ability to express themselves, saying, "Hi folks, mind if we join you?"

No, it's far easier to imagine *sapiens*, with greater numbers and superior technology -- they had throwable spears, but Neanderthals did not -- denying Neanderthals access to prime hunting grounds, like the bank of a river where animals congregated. They wouldn't even have to kill Neanderthals, not many anyhow, just force them to find food elsewhere, until they were driven away once again from that new habitat.

The greatest breakthrough in research about Neanderthals was the sequencing of their genome, for which the 2022 Nobel Prize in physiology or medicine was awarded to Svante Pääbo, a longtime Neanderthal scientist and a director of the Max Planck Institute for Evolutionary Anthropology in Leipzig, Germany. Pääbo was able to determine that the last common ancestor of Neanderthals and *sapiens* emerged about 800,000 years ago, before their lines branched off. His achievement has also opened the door to greater insight into how the physiology of modern humans compares with that of our ancestors.

Perhaps most surprising was the discovery that modern Europeans have about 1% to 2% Neanderthal DNA in their genome, and up to 4% in the genome of some Asians. So we know now that there was interbreeding between Neanderthals and *sapiens*. And the fact that most native sub-Saharan people have no Neanderthal DNA tells us that interbreeding occurred after *sapiens* migrated out of Africa.

Paleoanthropologists often mention such interbreeding without explaining exactly how it happened, that is to say, who fucked whom? For traces of Neanderthal DNA to be carried down the descent of *sapiens*, the offspring of that interbreeding, with its genes split 50/50 between the two species, would have to live with *sapiens* and then breed with *sapiens*, as would their offspring for many generations, in time diluting the DNA contributed from the Neanderthal side.

Given the territorial nature of *sapiens*, it is impossible for me to imagine a Neanderthal male wandering into a *sapiens* camp and boinking one of its females. It seems far more likely that *sapiens* abducted Neanderthal women, who would make valuable breeding stock. We've seen such behavior in recent human history. With fewer remaining females, this would reduce the capacity of small Neanderthal bands to reproduce. Even if other breeding females remained in those bands, it would further exacerbate the lack of genetic diversity often cited as a key reason for the Neanderthals' eventual demise.

By about 40,000 years ago, Neanderthals had disappeared from almost all of Europe, with a small population remaining in Spain. With their backs to the Mediterranean Sea but unable to cross to a warmer Africa, a period of deepening cold and changes to the food supply may have finally done them in. Neanderthals were neither as brutish as they had been depicted for so long nor as civilized as revisionist Neanderphiles would have them be. They were a successful species whose evolution stalled and whose small population groups suffered in encounters with *Homo sapiens*. They live on in our DNA.

An offshoot of their evolutionary line is known as the Denisovans. Their existence first came to light with the discovery of a finger bone of a female child in Denisova Cave in Siberia. A very few other Denisovan bones have been found at other sites, and recently a previously-known skull in China was newly identified as Denisovan by its DNA. In fact, what we know about Denisovans is almost entirely from DNA analysis, showing the growing importance of this powerful tool for paleoanthropology. One bone so analyzed proved to be the first generation of a Denisovan father and a Neanderthal mother. Either this was an extremely lucky find or, perhaps more likely, interbreeding among the two groups was not uncommon. So little is known about the Denisovans that they have not, at least as yet, been designated a separate species.

By the late Stone Age, "a fundamental behavioral transformation was taking place in *Homo sapiens* and sparking a revolution in the way in which hominins did business in the world," said paleoanthropologist Ian Tattersall, curator emeritus with the American Museum of Natural History, speaking in 2023 at a conference of the Institute of Human Origins.

BRAIN POWER

The key to how we've transformed is how our brains have evolved, but we have no direct way to measure those changes. Dead brains decompose. What we do have are the skull cavities of many ancient hominins. They increased in size over time, from about one-third that of modern humans, as was Lucy's, to the Neanderthals, who had skulls even larger than ours. It's now possible to make digital images of the inside of skulls, called virtual endocasts, to model what such brains looked like, and compare them to a modern brain, seen here.

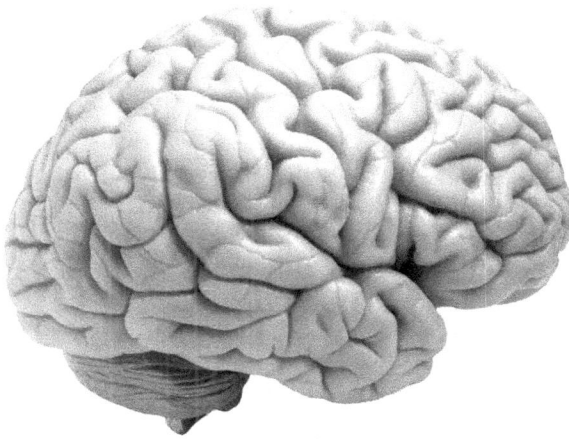

The key difference is the frontal lobe, on the right, and especially the pre-frontal cortex, a layer on the right of the frontal lobe. It handles higher-order cognitive functions such as language, decision-making and abstract thinking. The sloped foreheads of archaic hominin skulls meant they had smaller frontal lobes and were less developed cognitively than *sapiens*. The high-domed shape of our skulls, and the brains they enclose, is what characterizes us as modern humans.

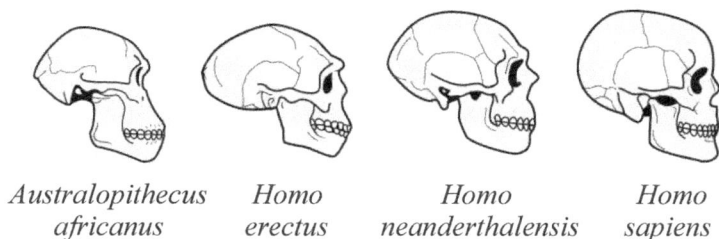

Comparative shapes of the skull (not to scale)

| *Australopithecus africanus* | *Homo erectus* | *Homo neanderthalensis* | *Homo sapiens* |

Oddly enough, there is evidence that about three to five thousand years ago, human brains apparently began to get somewhat smaller. No, we were not dumbing down, and just why this occurred is unknown. It may have been due to natural selection. As brains and skulls got bigger, child birth became more difficult, with rising rates of infant and maternal mortality. So nature selected for smaller but more efficient brains. Bigger is not always better.

Neanderthals had the biggest skulls of all. This is yet another reason to conclude that interbreeding was not between Neanderthal men and *sapiens* women. If the fetuses resulting from such matings had bigger heads and broader shoulders than those typical of *sapiens*, heaven help those women. On the other hand, with their stocky builds and wide pelvises, Neanderthal women had large birth canals, and would make good breeders for *sapiens* males. They could easily deliver their babies, at a time when producing offspring was critical for the survival of the tribe.

Whether a trend toward smaller but more efficient brains will continue, no one can say. But we can say that as humans evolved, more than any other trait our brains best defined who we were, and who we are.

SOLE SURVIVOR

Many species of our genus *Homo* have come and gone. Several lived contemporaneously with *sapiens*, if not in geographic proximity. One by one they all went extinct by about 40,000 years ago, except *sapiens*, which is long believed to have emerged about 200,000 years ago. But some fossils found in 2017 in Jebel Irhoud in Morocco cast that in doubt.

At first thought to be Neanderthal, the remains were later said to be *sapiens*, which would extend our origins to more than 300,000 years ago. Not everyone agreed, including renowned British physical anthropologist Christopher Stringer. The Jebel Irhoud skulls had a vertical face and modern teeth, but also pronounced brow ridges. They lacked the characteristic high rounded shape that differentiates anatomically modern humans from hominins with sloping egg-shaped skulls, like Neanderthals. The Moroccans were archaic humans in the process of evolving into *sapiens*, but not yet there.

Stringer is best known for advocating the "Out of Africa" theory, which says humans left Africa about 60,000 or so years ago, spreading around the globe and displacing other hominins. In *Sapiens: A Brief History of Humankind*, (Harper Perennial, 2018), Juval Noah Harari speculates that about 70,000 years ago, the species underwent a "Cognitive Revolution," a rapid neurological evolution that changed them from beings just living to survive, into thinking and feeling and talking modern people. Then they went off to see the world, and to conquer it.

Conventional wisdom now says that *sapiens* is the sole surviving species of *Homo* alive today. But is it?

Homo habilis

Homo sapiens?

Our skulls and bones may not differ anatomically from our ancestors dating back tens of thousands of years. Yet as paleo-anthropologist Ian Tattersall said of an earlier time, a fundamental transformation is taking place in how we do business in the world. Humans today think and behave in radically new ways, with radical new adaptations to old ways. For millions of years, our ancestors could only communicate interactively with one another face-to-face, or if they shouted loud enough. Now much of our interacting with other people is done remotely with electronic devices. No need to shout.

This was brought home in a dramatic way during the COVID-19 pandemic. Most of us had probably never been on a Zoom session before we were quarantined, but we did then and we continue doing so. There is an ongoing debate about remote work and remote schooling. But the fact remains that remoting is possible, and as we get better at it, we'll inevitably do more of it. Humans have always improved upon their tools, only abandoning them after coming up with something better, like when metal tools replaced stone ones.

We all know how fast change happens now, but consider this. It was 700,000 years or so between hominins first using Oldowan stone tools and then devising Acheulean ones. From the invention of dynamite by Alfred Nobel in 1866, to the detonation of an atomic bomb in 1945, was a mere 79 years.

Since we first learned to control electricity, our uses of it have vastly proliferated. We are undergoing profound changes in who we are and how we live. Humans evolved in the past, enabled by fire. Humans are still evolving, empowered by electricity. A new species is emerging. Call it *Homo electric*. *Homo sapiens* is no longer alone.

Poster from the U.S. Rural Electrification Administration, 1930's

PART FOUR

The Electric Age

My mother was determined that I was going to leave the farm and do well in life. She got a washing machine in 1942 as soon as we got electricity and she took in washing. She washed the school teacher's clothes and anybody she could and sent me for singing lessons for $3 per lesson.

-- *Johnny Cash, musician*

Twenty-mule team in Death Valley

RMS Titanic

FROM STONE TO STEAM

After lasting for some three million years, the Stone Age finally came to an end about 4,000 to 6,000 years ago. By about 12,000 years ago, humans began to abandon their hunter-gatherer ways to settle in larger communities and adopt an agrarian life. In the early days of farming they still used some stone tools, as well as sticks for digging. Not very productive. But then they learned to use metals – copper, bronze and iron – to make plows, other tools and weapons.

Their villages grew into small cities and independent states, then into large cities and nations. This was made possible by the unique human ability for large-scale cooperation and by industrial, technological and financial innovation. They developed complex languages, invented many useful tools, created literature, art and music, and waged ever more deadly wars. Eventually *Homo sapiens* formed new societies governed not by fear and brute force but, however imperfectly, by the consent of the governed. An Age of Enlightenment culminated in the founding of the United States.

But it was still a mechanical world. Muscle power – human and animal – and natural forces – wind, water and fire – powered tools and machines of increasing complexity, such as printing presses, sailing ships and lumber mills. This had its limitations, depending on the availability of local resources like flowing water to turn a mill wheel. There was no water to operate a mill to process borax salt in California's Death Valley, so large wagons of borax had to be hauled elsewhere to be processed. But you could only hook up so many mules to a wagon, like the famed teams of twenty.

This dependence on local resources changed dramatically with the invention of steam engines, which could be operated almost anywhere. Initially they were conceived for pumping water, like the ones put into commercial use in 1698. The first steam engine to transmit power to a machine soon followed in 1712. Scotsman James Watt introduced an improved version in 1776, laying the basis that fateful year for another revolution, an industrial one. The first commercially successful steamboat was built by Robert Fulton in 1807, using a Watt engine. Soon steamships were crossing the oceans, carrying more passengers and cargo faster than had ever been possible in the heyday of sail. On land, steam powered the trains which traversed the United States, Great Britain and elsewhere.

The steam engine is often depicted as a great leap forward. But just how revolutionary was that "Industrial Revolution"? Steam power greatly increased industrial output and the speed and capacity of transportation. But a steam engine was essentially a fire in a box fed fuel by the muscles of men, the culmination of the use of fire, not the true revolution soon to come.

The greatest ocean liner yet built was launched in 1912. It needed 176 "firemen" shoveling half a ton of coal every minute into 159 furnaces to heat 29 boilers to produce the steam which drove the pistons which turned its three massive propellers. It didn't seem like much of a revolution to those men, amid the sweltering heat and choking coal dust. Most of those poor souls went down with their ship when the RMS *Titanic* sank. They had depended on eyes and ears which failed to steer a safe course past the looming iceberg. Sonar had not been invented yet. The first patent for an echolocating device was filed in England a month after the sinking.

THE GOLDEN SPARK

As some men tinkered with steam engines to find a new source of power, others were trying to solve the problem of how to communicate back-and-forth beyond the sound of the human voice. Systems for signaling go back as far as 400 BC, when messages were sent from the towers along the great wall of China using fires, flags and eventually gunshots. Signal fires were used extensively by the Roman military, and smoke signals by indigenous Americans and aboriginal Australians. But all these methods were limited to predefined meanings.

In ancient empires like China, Rome, Egypt and Greece, couriers carried messages to and from the elite few, but not the common folk. The Incas had an elaborate network of couriers who ran in relay teams over enormous distances in the rugged terrain of the Andes. They delivered information for the ruling class using a coded system of knots on strings called *quipu*. Around the time Peru was conquered by the Spanish, five hundred years ago, King Henry VIII of England established the Royal Mail, but it was over a century before commoners were allowed to use it. Of course, most people could not read or write, and needed someone to help them.

In the American colonies, a postal service was established just before the Revolutionary War. The first U.S. Postmaster General, Benjamin Franklin, sped up mail delivery between New York and Philadelphia by running carriages day and night with relay teams of horses. There was much to read and write about in those days, for those who could. The desire to send and receive mail contributed to the proliferation of school houses to teach reading and 'riting and 'rithmetic too.

By the mid-18th century, inventors experimented with optical telegraphs, which used visual signals from shutters or paddles to spell out messages, although very slowly. But by the early 19th century, attention was focused on using electricity to operate telegraphs, at first primarily for signaling along railroad lines, as was done in England in 1837. But that complex five-wire system was too costly for general use.

Then in 1844 a spark of innovation sent an impulse from Washington over a single wire to make a device at the other end in Baltimore quiver with the "dits" and "dahs" of an encoded message. It was the first long-distance test of the telegraph invented by Samuel F.B. Morse. Using his eponymous code, a new form of language spoken by fingers not mouths, he asked the question: "What hath God wrought?" Had I been there, I would have replied, "The Electric Age."

The telegraph was the first widespread commercial use of electricity, a disruptive technology. In the spring of 1860, the Pony Express began mail delivery from Missouri to California in only ten days, what was then a remarkable feat. To do it, riders switched horses stationed along the route, like Ben Franklin's mail carriages. Yet just a year and a half later, when a telegraph link was completed along the Pony Express route, they put the ponies back in their stables and closed the doors on the service. Electricity was faster than horses.

The telegraph electrified the American public, and sparked the imaginations of writers and thinkers. In *Telegraphies: Indigeneity, Identity and Nation in America's Nineteenth-Century Virtual Realm* (Oxford University Press, 2019), Kay Yandell wrote that Henry David Thoreau, of Walden Pond fame, said telegraph wires "told of things...worthy of the

electric fluid," not only "the price of cotton and flour, but …
hinted of the price of the world itself and of things which are
priceless, of absolute truth and beauty." She noted that rags-
to-riches storyteller Horatio Alger "in his popular book *The
Telegraph Boy* (1879) enacts the American Romance through
telegraphic visions of the American Dream."

Yandell cited Mark Twain seriously "postulating that a
'mental telegraphy' operated daily between the unconscious
minds of Americans." Abraham Lincoln sent nearly 1,000 telegrams, which he called "lightning messages," to his generals
during the Civil War. The "Great Emancipator" was a great
communicator. Electricity helped win the war to end slavery.

The driving of a golden spike, completing a transcontinental railroad, is remembered as a pivotal historic event. But it was the golden spark of the telegraph, vital to operating the railroad, which connected the nation through a system of wires, the first wide area network. It was soon followed by another network, for transmitting the human voice: the telephone. For the first time you could have a conversation with someone from afar, no longer needing to be within earshot.

Today of course we're connected with each other in our homes, schools, workplaces, everywhere, with access to an ever-growing array of information and services. From birth, we are nodes on networks wired and wireless. Electricity is what makes it all possible; it's the foundation of modern civilization. We depend on it, we expect it always to be there, we want the things it brings us, we cannot live long without it.

Some people call the time since 1945 "The Nuclear Age" or "The Atomic Age," because of our ability to unleash massive destruction with nuclear weapons. But without electricity, there would have been no way to develop, deliver and detonate atomic bombs over Japan. For better or worse.

Some people call the time in which we live "The Digital Age" or "The Information Age," because of the importance and pervasiveness of computing. But without electricity, there would be no computers, no smartphones. Without our ability to control electricity, there would not be much of anything as we now know it. Our lives would still be based on fire.

No, it is "The Electric Age" we live in. Since learning to control electricity, humanity has been undergoing a profound transformation. We are evolving beyond *Homo sapiens*. We are becoming a new species, *Homo electric*.

ELECTRIC FIRE

Benjamin Franklin was one of the most remarkable men of his time. No gentleman farmer-slaveowner was he, but an urbane man of many interests, talents and accomplishments: writer, publisher, political leader, diplomat, inventor and innovator. It was as a scientist that he became fascinated with electricity. Franklin was the first to term electrical charges as positive and negative. On Thanksgiving you might try his method of killing a turkey with an electric shock and roasting it on an electric spit. He claimed the result was "uncommonly tender."

Franklin is famed for his experiment flying a kite to prove that lightning was electricity. He did not do it during an electrical storm and wait for it to be struck by lightning, a dangerous activity. Instead, he flew a kite into a storm cloud to gather an electrical charge. As he described it, "When rain has wet the kite twine so that it can conduct the electric fire freely, you will find it streams out plentifully from the key at the approach of your knuckle." Fire in a wire.

Michael Faraday

Westinghouse Electric Company ad for AC

The renowned English scientist Michael Faraday, born in 1791, a year after Franklin's death, was a great admirer of Ben. Like Franklin, Faraday was a man of many talents, but he is best known for his work with electromagnetism. He was the first to demonstrate how to convert electrical energy into motion, the basis of the electric motor. In 1831 he discovered how a magnetic field could induce an electrical current, which led to the invention of the electric generator.

When the generator became commercially feasible later in the 19th century, companies were launched to produce and sell electricity. The Edison Electric Light Company (it later became General Electric) was founded by Thomas Edison to supply power for the incandescent bulbs he had invented. Its chief competitor was also named for its founder, George Westinghouse, who teamed with electric wizard Nikola Tesla.

They engaged in a "war of the currents." Edison provided direct current (DC), which was low power and could only be distributed a half-mile or so from the generator. Westinghouse sold alternating current (AC), using electricity supply systems invented by Tesla. He hired an engineer named William Stanley Jr., who built the first working electrical transformer. It enabled high-voltage AC power to be transmitted over long distances and then stepped down by the transformer to provide the lower voltages used for lights and other on-site applications. In 1886 Stanley demonstrated the first full AC power system, lighting offices and stores on Main Street in the small western Massachusetts town of Great Barrington.

The transformer was, well, transformational. Stanley and Tesla made possible the large-scale grids for electricity distribution which we rely on today.

WHAT ELECTRICITY DOES

Electricity *powers* things, whether a light bulb or a car. It *multiplies* the force of human muscles, whether with an electric drill or a fork lift. It *magnifies*, enabling us to see into the atom and deep into space-time. With computers it *extends* the capability of the human mind, beyond the limitations of our unaided brains, to store, access and process information. It is driving human evolution by *connecting* people, *amplifying* those connections over distance and time, and *accelerating* adoption of new techniques, technologies and behaviors that characterize who we are and how we live.

For millennia, people didn't travel more than about 30 miles from their birthplace and had contact with few others. Since the onset of electricity, people's interactions with and awareness of one another have changed radically, from limited, face-to-face, and local, to extensive, often remote, and even global. We're not the shepherds we once were, tending our flocks in pastoral isolation, but are inseparable members of the human flock interconnected by electricity.

We share with our forebears of old many of the traits, good or bad, which make us human, and we likely always will. The words of Shakespeare are immortal, but the people reading and hearing them today are very different from those who watched his plays at the old Globe Theatre. From an evolutionary perspective, they were more closely related to those *Homo sapiens* who first settled down as farmers 12,000 years ago than to the beings we are now becoming, *Homo electrics*. It has been estimated that humans of 200 years ago, before electricity, had an average IQ, based on today's scale, of 70.

With electronic devices and social media, our connections with one another are becoming pervasive and always-on, more often remote than in person. Do people misuse those devices and media? Yes, unfortunately. We must learn to be smarter in how we use them. But to lament that people are becoming addicted to them is a misplaced notion, just as it would be to say that we're addicted to electricity, or air. Sure, you can disconnect for an hour or a day, go camping in the woods for a week, or live in a remote place like the Andes for months. But sooner or later you will choose to return to the electrified world. It's our natural habitat. You want to live as an *electric*, because you are an *electric*. Our electrified connections and devices are inseparably part of who we are.

Facebook, launched in 2003, experienced exponential growth, like other social media. Over three billion people are now monthly active users, two billion daily active users. This is not due solely to any "genius" which might be attributed to Mark Zuckerberg. He created the software with several Harvard friends as a way for students at the college to rate how hot-or-not their classmates were. Previously, students had checked out head shots in an annual print publication issued by Harvard called The Facebook, and shared snarky comments, maybe over a beer at a local pub.

But what Zuckerberg inadvertently tapped into, and to be sure relentlessly and successfully exploited, is the desire of *electrics* to be connected with one another, a desire amplified by applications such as Facebook. The satisfaction of this desire has been accelerated by the widespread adoption of computers, smartphones, the internet and social media, the characteristic new human tools of the Electric Age.

IS IT REAL, OR IS IT...?

It needs to be asked whether our electronic interactions can convey the same emotional weight as those face-to-face. In the 1980s, a manufacturer of cassette tapes called Memorex, to promote the quality of recordings made with its products, used the advertising slogan "Is it real, or is it Memorex?" Can we likewise distinguish remote communications as somehow less "real" than those in person? I'm sure you have had, for example, phone conversations or email exchanges that were meaningful to you, and emotional: joyous, sad, funny, angry.

There is a long (and sometimes tragic) history of scientists experimenting on themselves when no other subjects were available. I was the involuntary subject of an "experiment" of my own, living for two years almost exclusively with remote communications. Two weeks after the first COVID-19 stay-at-home order in 2020, my wife Jan passed away, after a 30-year battle with breast cancer. At the time we had been living physically remote from neighbors deep in the woods of western Massachusetts. Because I was immunocompromised and particularly vulnerable to the virus, I had to carefully limit my contact with other people.

I went out about once a week to do necessary errands like food shopping (there was no delivery available to my house), wearing a mask of course. But other than a brief exchange with the cashier ("Did you find everything you were looking for?") and on other rare occasions, for two years I had almost no face-to-face contact with people. No funeral or memorial service. No friends and family visiting to offer condolences and comfort. Some neighbors did stop by with homemade meals (chicken soup!), but they left it outside the door after a brief conversation standing far apart on the driveway. No one came in the house, except a repairman once when my furnace went on the blink in the dead of winter. I kept my distance.

Yet I did have regular contact with friends and family remotely by telephone, email, text, Facetime and Zoom. Did these replace physical interaction? Yes and no. No one could hug me at a time when I badly needed it. But remoting was a lifeline it's impossible for me to imagine not having under those circumstances. And fortunately, installations of fiber optic connections in my town had begun shortly before the pandemic struck, and arrived at my house not long after.

I found that even in writing an email, I could convey my feelings when I concentrated on the person I was sending it to. Yes, there were plenty of quick texts and the like which I dashed off without feeling. But when I summoned my feelings they came through to the person on the other end, as did their communications to me. People did just that for centuries when writing letters. Imagination is powerful.

Today humans have numerous electric-powered facilities to convey and receive emotions. We have a strong need to be connected with one another so we can share our feelings. Electricity is making that feasible under circumstances when face-to-face contact is not possible, and frequently replacing in-person contact even when it is possible. Much concern has been expressed about how young people use their devices and social media for negative purposes like shaming, and withholding contact by ghosting. These are ages-old social behaviors expressed and amplified through powerful new means. Still, we need to use these tools better and wiser.

WHO ARE WE?

The evolution of humans from *sapiens* to *electric* is real, ongoing and accelerating. We're experiencing profound changes which raise questions about our future. How will we govern ourselves and coexist with our fellow humans? How will we survive political unrest, pandemics and climate change? How will we advance human rights and protect the ecosystem of our home, the planet Earth?

Electrics network. We adopt new ways to connect with each other, to learn and to act through online collaboration. We use applications such as Zoom to form and develop relationships with people we haven't met face-to-face, and perhaps never will. We watch remote presentations by people who may not even know we are there, or who we are. We join virtual communities which transcend distance and national borders, connecting us to people with similar interests and beliefs. Sometimes those communities gather in the real world, to act out as members of our collectively self-defined tribes, at Burning Man and Comic-Con.

We decide who we are and want to be on the spectrum of gender. We create our public personas by the clothing and symbols we display on our bodies, and the words and images we post online. In the game of life in the Electric Age, we are our own avatars. Even the money we exchange is increasingly electric. We mark time with electric devices.

Some of the changes we are undergoing are invisible and as yet unknown. What are the long-term effects on our brains, nervous systems, organs and genes from being surrounded and penetrated by omnipresent electromagnetic energy and its applications like radio, TV, microwave, cell phones and wi-fi? The relentless increase in computing power, especially artificial intelligence, raises questions about how we will develop and control it, or perhaps be controlled by it.

Why our evolution is happening is a question of both science and belief. We don't yet understand how our brains changed from those first *Homo sapiens* 200,000 years ago. Perhaps there really was a Cognitive Revolution, as Harari proposes, which transformed human thought. Something did.

Now we are undergoing an Electric Revolution. Could we imagine a scenario in which humans harnessed electricity to provide light and heat and to power mechanical devices, but stopped development there? No, that's inconceivable. It's inevitable as humans that we would push its uses further and deeper and wider, changing us in the process. Electricity is empowering humans to evolve beyond *Homo sapiens*.

Why, you might ask, does it matter whether we call our species *Homo sapiens* or *Homo electric*? It should, because how we see ourselves and our relationship to the world around us, what we believe, affects how we live, act and think in that world. Identifying as *Homo electric* is part of the process of evolving beyond *sapiens* to *electric*. Of accepting that who we once were as humans we no longer can continue to be if we are to survive. Of outgrowing the limitations of our ancestors, the territoriality that divided them, too often by intraspecies violence. Of embracing awareness that we are all part of an interconnected planetary ecosystem, a web of life. Of realizing that who we are becoming today is but a step on a long and unfolding evolutionary path leading us onward toward who we will become tomorrow, on a Planet Electric.

"I know things Google doesn't."

A Word From The Wise

Speaking of who we will be, let me digress briefly with a story. A brilliant but homely math professor is considering whether to marry a beautiful woman who is notably lacking in the smarts department. He thinks, "Imagine if our children had my intelligence and her looks!" Gaming out potential scenarios, he wonders, "But what if they had my looks and her intelligence?" After many fruitless hours on Google trying to find the answer to his quandary, he decides to seek advice from the Wise Old Man who lives atop a nearby mountain.

After struggling for hours climbing the rugged peak, the professor reaches the top and finds the Wise Old Man at home in his aerie. Seeing the professor clambering over the rocks, the Wise Old Man says to him, "Welcome, my son. How can I be of help to you?" The professor, somewhat out of breath after his climb, replies, "I'm seeking some answers about heredity." Smiling, the Wise Old Man says, "You've come to the right place, my son. I know things Google doesn't."

"So tell me," the professor says, "will my children have my brains and her looks, or her brains and my looks?" To which The Old Man replies, "It depends." Somewhat irritated by this vague answer after his struggle up the mountain, the professor presses his point. "Depends on what?"

The Old Man pauses for a moment, and then with a beatific smile crossing his face, says "On everything."

PART FIVE

The Evolution of "Evolution"

I think that all the years of exposure to amps and electricity has altered my body chemistry.

-- Iggy Pop, punk rocker

Caricature of Charles Darwin
after the publication of The Descent of Man

THE ORIGINS OF EVOLUTIONARY THEORY

Since this book is about human evolution, before going any further let's consider what we're talking about when we say "evolution." It's a word with a very specific history and meaning, but a meaning which has itself begun to evolve.

You're probably familiar with a version of the illustration a couple of pages back, what is sometimes called "the march of progress": from a monkey shuffling along the ground many millions of years ago to an early human first walking upright to *Homo sapiens* striding along confidently. It's a representation of genetic evolution, the physiological changes (most of which are not visible) that mark the emergence of a new species. It's a symbolic but not a literal depiction of evolution – the human figure did not descend from the Neanderthal behind her – but you get the idea. As Darwin made clear in *The Descent of Man*, we're descended from primates, and you've read how that evolution proceeded from the time when hominins and chimpanzees branched off.

It's worth repeating the basic elements of Darwin's theory of evolution. "Many more individuals of each species are born than can possibly survive," and in "a frequently recurring struggle for existence" any being with a variation of a trait however slightly advantageous "will have a better chance of surviving, and thus be *naturally selected*."

Natural selection became popularly known as the "survival of the fittest." The phrase was not coined by Darwin but by the philosopher Herbert Spencer after reading *Origin*, then Darwin adopted it. His point was not pitting one being against another, like contestants in an episode of "Survivor!" Rather

Gregor Mendel

it was the *process* by which, given the natural variance in how a trait manifested itself in different individuals, a more favorable version of the trait would be selected, passed down to offspring and then disseminated throughout a species.

This occurred as the result of what Darwin called "the strong principle of inheritance," whereby "any selected variety will tend to propagate its new and modified form." So through variance, natural selection and heredity, enough new traits could be accumulated in a species to cause it to evolve eventually into a new one.

Keep in mind that "species" is a descriptive scientific term, part of that taxonomy mentioned earlier. A new species can be said to have evolved only when scientists proclaim it to be a new species. Members of that species may be doing better than their forebears, surviving fitter as it were, yet remain blissfully unaware that they have become a new species. Like people who, if they think about it at all, may be under the impression that they are *Homo sapiens*, when they no longer are.

Darwin may have believed in the principle of inheritance, critical to his theory of evolution, but neither he nor any of his scientific contemporaries in England understood the mechanism behind it. Unbeknownst to them, at that very same time an obscure monk in what is now the Czech Republic was conducting experiments about the inheritance of traits, using pea plants he grew in the garden of his monastery.

Gregor Johann Mendel had grown up on a family farm, so he was familiar with the crossbreeding of animals to improve the stock, and was also exposed to the subject of animal heredity at the university he attended. At the monastery he chose peas to be the subject of his research because they reproduced quickly and exhibited several identifiable traits which he could easily track, such as seed shape and flower color. From 1856 to 1863 he cultivated thousands of pea plants to test his hypotheses about inheritance.

From this work he formulated what are known as Mendel's Laws of Inheritance, how dominant and recessive "factors" (they were not yet called genes) produced traits in crossbred offspring. He presented a paper on his findings in two lectures in 1865; it was published in 1866. Unlike Darwin's *Origin*, which achieved great acclaim and popularity immediately upon its publication, Mendel's work was misunderstood and ignored, forgotten when he died in 1884. It was not until the early 20th century when his paper was rediscovered, his experiments were repeated and verified, and "Mendelian inheritance" was accepted as scientific doctrine. The term "gene" was introduced in 1905 as the inheritable unit for the transmission of biological traits, and Mendel was dubbed retroactively as "the father of genetics."

Neither Darwin nor Mendel knew anything of each other's work. For decades they were two distinct lines of thought on separate scientific tracks. It was not until 1930 that the re-nowned British statistician Ronald Fisher brought them together in his book aptly titled *The Genetical Theory of Natural Selection*. He argued that Mendelism validated Darwinism, in that it explained the process whereby the results of natural selection were transmitted to successive generations. This merger of the two in a mathematical framework developed by Fisher and others became known as the "modern synthesis" of evolutionary theory.

While genes were thought to be responsible for the inheritance of traits, how that process worked was not well understood until 1953, when James Watson and Francis Crick discovered DNA (deoxyribonucleic acid), explained its role in the transmission of the biological instructions which govern heredity, and identified its intertwined double-helix structure.

It could be said that they "inherited" that helical image of DNA from Rosalind Franklin. Two years earlier she captured its two strands in an X-ray photograph known as "Photo 51" (below). It may not look like much, but when Watson saw it, he said "my mouth fell open and my pulse began to race." It has been called the most important photo ever taken, a microcosmic genetic mandala. But Franklin died before getting the recognition due her. Watson and Crick got the Nobel Prize.

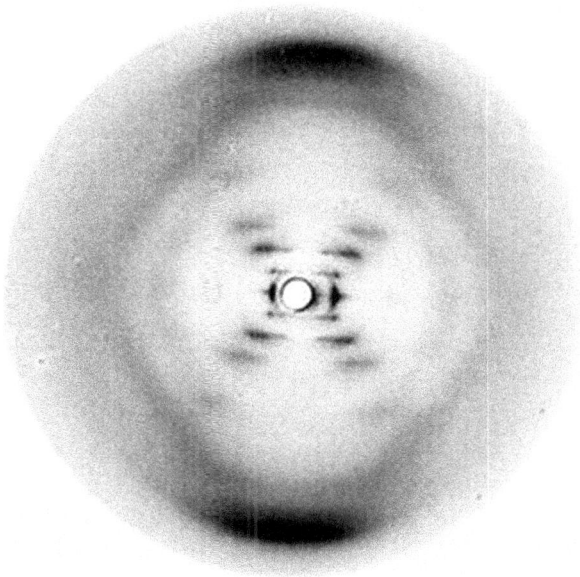

IS IT JUST ABOUT THE GENES?

Variation. Natural selection. Heredity. DNA. These are the legs on which modern evolutionary theory stands. Key to it all are the genes, segments of strands of DNA which are copied and recombined at conception, half from the mother and half from the father. They determine the traits which are selected by nature and passed along from one generation to the next.

For a few theorists the gene is everything, most notably Richard Dawkins, author of *The Selfish Gene* (Oxford University Press, 1976). To these genophiles, natural selection does not work on organisms but on genes themselves. It is a specific gene and its counterparts in other organisms which are struggling to survive and pass on copies of themselves. What matters is the global population of that gene, not its existence in particular organisms, like you or me. To Dawkins, we are merely "survival machines" for genes, convenient containers of skin and bones. Genes build them, exploit them and ultimately discard them when the organisms die. But copies of their genes live on in subsequent generations of those organisms. In Dawkins's view, it is the fittest genes which survive.

Most evolutionary scientists, however, do not take such an extreme view about genes. It is the survival of the organism that is the key to evolution, as Darwin said. If anything, the importance of genes as absolutely controlling heredity and thus evolution has been diminishing with recent advances in evolutionary thinking.

It's interesting how many key advances have occurred in roughly half-century cycles. Darwin's *On the Origin of Species* was published in the mid-1800s. Mendel's work on heredity was done around the same time, but only surfaced publicly in the early 1900s. It was in the 1950s that Watson and Crick revealed the nature of DNA. And with the dawn of the 21st century, ideas began to emerge which challenged the sole role of the gene in determining heredity, as we shall see.

Back in Darwin's day there was a school of thought called Lamarckism, named after its principal proponent, that traits acquired during your lifetime could be passed on to your descendants, like, say, muscles gained from hard physical work. Even Darwin himself allowed that this might be possible. That notion, however, was banished from evolutionary thinking in favor of what has long been the unquestioned theory that genes alone control heredity, and that characteristics acquired by an organism during its lifetime die with that organism and cannot be passed on to its descendants. You'll have to build those muscles yourself.

Heredity was said to operate like a blueprint, a fixed plan for combining the genes of father and mother to produce offspring whose characteristics are predetermined by those genes, like Mendel's peas. But we are now beginning to understand that factors outside genes impact how genes behave, or as the geneticists say, how genes express themselves. The Greek letter $\sum$ (sigma) is used by mathematicians, like our homely professor, to symbolize the summing of things. What I will call the "sigmasphere" is the sum total of everything outside the genes themselves which can act on and affect gene expression, and thus affect heredity and evolution, as I will explain. While genes remain stable over many generations, the actions of the sigmasphere are giving rise to a new species, *Homo electric*.

INHERITANCE THROUGH EPIGENETICS

"Epigenetics" is a term first coined in the 1940s but which has come into prominence in the 21st century. "Epi" is from the Greek for "above" or "on top of." In the realm of genetics, the prefix refers to molecules which interact with and operate on top of the genes, activating or deactivating them. For a group of genes, some of them turned on and some off, the effect can be like a dimmer switch, controlling how strongly or weakly those genes express themselves. Such molecules are referred to as "epigenetic marks" on the genes.

Epigenetics is the system which mediates how genes interact with the "contexts" in which they exist, and how that influences the development of our characteristics. Those contexts can include the biological but nongenetic aspects of the inside of the body; the external physical environment; and the human cultural environment in which we live. That is to say, the sigmasphere, the sum of all the influences on gene expression.

Just as an organism's full array of genes is known as its genome, so too does it have an epigenome. The genome remains constant from conception onward (except for the rare occurrence of gene mutations), but its epigenome does not. Characteristics acquired by an organism attributable to changes in the epigenome can be inherited with the epigenome and its epigenetic marks. It is not only our DNA – genetics – but our experiences in the sigmasphere acting on our DNA – epigenetics – which determines who we are.

While the genetic evolution of a species proceeds glacially, with little change from one generation to the next over many millennia, epigenetics can influence the evolution of a species even in the span of just one generation to another, due to epigenetic inheritability. In *The Developing Genome: An Introduction to Behavioral Epigenetics* (Oxford University Press, 2015), David S. Moore writes that "for anyone committed to the idea that only genetically determined characteristics can be inherited, a major upheaval is coming." He argues that people's characteristics, both physical and psychological, "develop from gene-environment interactions; none have a strictly 'genetic basis'." He calls the idea that only "hard" genetic inheritance can transmit traits "hopelessly simplistic...because genes do not operate in a vacuum." As I put it, they operate in the sigmasphere.

Moore cites an example of epigenetic inheritance of acquired characteristics involving laboratory rats, in research conducted at McGill University in Montreal. Rats have more in common with humans than you might like to think. Newborn rats, like our babies, can't fend for themselves and require a mother's care, and they have nervous systems quite similar to our own.

Mother rats lick and groom their babies, but to different degrees, a genetic trait. Some tend to be high licking-and-grooming (LG) mothers, others low LG. In the McGill experiment, low LG mothers (the first generation) had female pups (the second generation) which were placed soon after birth with high LG foster mothers, and grew up to become high LG mothers. Their female pups (the third generation) turned out to be high LG mothers, even though from a strictly genetic point of view they were descended from the second-generation mothers who were born with the trait for low LG.

But those mothers had acquired the trait of high LG from their foster mothers, and passed it on to their offspring.

Licking and grooming benefits rat pups by reducing vulnerability to stress, which is moderated by a protein in their brains. Pups with low LG mothers had reduced expression of the genes involved with producing that protein. Less protein, more stress. But being raised by a high LG mother increased expression of those genes. More protein, less stress. This pattern of gene expression is an inheritable acquired trait. This is epigenetics at work, a desirable acquired trait, high LG, being passed down to offspring.

Inheritance of advantageous traits is of course the key to evolution by natural selection in the modern synthesis of evolutionary theory, in which genes rule. But if traits can be acquired through life experiences, and their associated *patterns of gene expression* passed on epigenetically, then such experiences can and do influence heredity and evolution.

Moore concludes that no longer viewing genes as solely determining who we are should change our view of human nature. "We are all profoundly influenced by the contexts in which we develop, and we have some control over those contexts; therefore, it is our responsibility to do what we can to help ourselves and others grow into compassionate, enlightened and fulfilled individuals."

That ethos is well suited to the evolving species *Homo electric*, with our increasing understanding of the contexts in which we develop and live (the sigmasphere), a growing ability to control them through science and technology (based on electricity), and a more enlightened view of our place on the planet and our responsibilities to it and to all living beings.

CULTURAL EVOLUTION

We human beings are animals, products of evolution from our animal forebears, as Darwin observed in *The Descent of Man*. But we are animals with a critical distinction from our fellow creatures: we are *cultural* animals. As such beings, our culture impacts who we are and how we are continuing to evolve. It is part of the sigmasphere affecting gene expression.

By "culture" I do not mean merely the arts, music and literature, but rather in the much broader sense of the characteristics and functioning of a human society: its laws, norms, traditions, customs and moral values which govern the behavior of its members; its languages and other forms of communication; its tools, sciences and technologies; its body of common knowledge which children and adults can acquire and use as they develop and age; its forms of creative expression; its religious belief systems; its balance between competition and cooperation among its members, as well as in its relationships with other societies; and its capacity for discovery, innovation and change.

Other animals do not have such complex cultures. They may be organized into groups like herds and packs. They may be governed by rules of hierarchy and behavior. They may teach their offspring by example and punishment. They may, as chimps do, groom one another as a form of social bonding and use simple tools. They may even attach themselves to humans for survival and comfort, like the ancestors of cats and dogs did. But animal societies do not rise to anywhere near the level of complexity of human culture, and they remain largely static, whereas human culture continues to evolve.

Culture began to arise in our hominin ancestors as early as two million years ago. As bands of humans over many generations accumulated information which aided their survival, those individuals who were best able to utilize such information were most fit to survive. I mentioned earlier the notion of a "Cognitive Revolution" about 70,000 years ago which was posited by Yuval Noah Harari. It may well have occurred because of the collision between evolving culture, with its increasing body of useful practices and techniques, and the functioning of the brain. When that body of information reached a critical mass, it led to the genetic evolution of the brain by natural selection, which favored those individuals with brain power better able to process and make use of it.

In his book *The Secret of Our Success: How Culture is Driving Human Evolution, Domesticating Our Species, and Making Us Smarter* (Princeton University Press, 2016), Harvard professor Joseph Henrich says that for hundreds of thousands of years, cultural evolution has been the central force driving human genetic evolution. He calls the interaction of culture and genes "the culture-gene coevolution," which he says "drove our species down a novel evolutionary pathway not observed elsewhere in nature, making us very different from other species – a new kind of animal."

Certainly, one of the most significant developments in human cultural evolution which differentiated us from other species was the emergence of language. At first language had to conform to the limitations of the ape-like brains of our early ancestors, but then natural selection began driving genetic evolution of their minds and bodies to improve their communications abilities, which aided their survival.

Henrich details the effects of culture on that genetic evolution. For example, it "pushed down our larynx to widen our vocal range, freed up our tongues and improved their dexterity, [and] whitened the area around our irises (the sclera) to reveal our gaze direction," as well as endowing us with the capacity for using cues like eye contact and pointing. It also affected the structure of the brain, "from expanding our hippocampus [which is involved with memory, learning and emotion] to thickening our corpus callosum (which connects the two halves of our brains)."

Henrich says that since groups of humans began to live in settlements, competition between groups led to increasingly larger and more complex societies, driving cultural evolution and shaping us as social beings. "Our ability to learn from others, to generate greater technological sophistication and larger bodies of adaptive knowledge" has given rise to what he calls "our collective brains." This, he argues, more than individual invention, has been responsible for the success of our species. Larger and more socially interconnected groups have larger collective brains, generating even further cultural evolution.

As we become more interconnected in the Electric Age, through the internet and other electronic means, human culture will evolve rapidly, as will our collective brains, leading in time to genetic evolution. To rephrase Henrich, this culture-gene coevolution is driving humanity down a novel evolutionary pathway making us very different from other species of humans – a new kind of human, *Homo electric*.

David Sloan Wilson is an evolutionary biologist. In *Evolution for Everyone* (Bantam Dell, 2007), he too stressed the importance of cultural evolution. "The primary human adaptation… is for our behaviors to be acquired less and less directly from our genes and more and more from other people." We share what we learn, thereby we evolve. We are crowdsourcing our evolution.

Wilson contrasted human evolution with that of social insects like ants and bees: "there are many thousands of species of social insects… [but] we achieved worldwide ecological domination while remaining a single species. The reason is that our diversification was cultural rather than genetic." There have been thousands of human cultures around the world, yet unlike insects, humans all have much the same genetic makeup. We achieved our cultural diversification in thousands of years, whereas a comparable genetic diversification would have taken millions of years. The reason, Wilson says, is because "our capacity for culture shifted evolution into overdrive."

He cites two tribes in the upper Nile basin of Africa long studied by anthropologists. The Nuer were originally a sub-tribe of the Dinka, then due to cultural differentiation became a separate tribe, and then the dominant one by absorbing many Dinka by marriage. Tribal members were no more consciously aware of this as cultural evolution – it was just how things were – than they would have been of genetic evolution.

Homo electrics, on the other hand, are very aware of how our electric culture is evolving and actively participate in that process, individually and collaboratively. In *This View of Life: Completing The Darwinian Revolution* (Pantheon, 2019), Wilson said we must be "consciously evolving our collective future." Because we can. Because we must.

THE ENVIRONMENT

Since the planet Earth was formed over 4.5 billion years ago, until about one hundred years ago, the only significant manifestation of the electromagnetic spectrum in its atmosphere was from the earth's inherent geomagnetic field, and from the sun and stars. Then humans began to harness it for communications. Again, telegraphy was at the forefront of change, when Guglielmo Marconi invented the wireless telegraph, which used electromagnetic radio waves. In 1897, demonstrating for the British government how it worked over open seas, he sent the message, "Are you ready?"

Were we ever! By the 1920s people could tune their radio sets to numerous stations (frequencies on the electromagnetic spectrum) that were "on the air," because their radio waves were being transmitted everywhere in the air. By the 1940s, television signals (higher-frequency radio waves) were becoming available. Change the channel (the frequency) to receive a different transmission. By the 1970s, big backyard dishes could receive TV signals beamed from satellites in fixed geosynchronous orbit 22,236 miles above earth.

In the 1990s, cellphone networks proliferated. You could keep talking in a moving car as your connection was handed off from the coverage area of one tower to another, often connected by microwave links. By the early 21st century, people in rural areas without wired internet connections could go online via high-orbit satellites serving huge geographic areas. It wasn't great, but it worked. Now a new generation of low earth orbit (LEO) internet satellites is being deployed by SpaceX, Amazon and others. There will be tens of thousands of them at an altitude of several hundred miles, making wireless internet widely available by sending signals back and forth to millions of users.

For those of us living in the developed world, our bodies and brains are continuously bombarded by electromagnetic radiation (EMR) from more and more sources. Many life forms live immersed in water. We live immersed in EMR, a product of our cultural evolution. It is as if the classic image of early evolution – creatures crawling from the sea onto dry land and adapting to life there – is playing in reverse, and we are crawling from dry land into a sea of EMR.

How are we adapting? The effects on human genetics (and those of all living things) of existing in such an environment are as yet unknown. It would be interesting to compare the genomes of some contemporary humans with those from some people who lived two hundred years ago, before electricity, to see if any differences have shown up... yet.

To date, public focus has been not on evolution but on the possible health effects of, say, living close to cell towers. There is no clear evidence of what such effects might be, and if you're concerned about it and are able to, you can move. But few people will want to move away from someplace where they can get good coverage for cellphone and television (cable TV systems use large dishes to receive those signals before sending them to you over a wired connection). Good coverage means that TV and cellular signals are permeating the atmosphere from broadcasters, cell towers and satellites.

The proliferation of EMR produced by human activity is a product of the evolution from *Homo sapiens* to *Homo electric*. Powering all those devices emitting EMR is, of course, electricity. Once it is reliably available in a particular area, electricity becomes a permanent feature of how people there live in succeeding generations, a defining trait of their culture. Electrification of a society – a village, a city, a country – is as much an evolutionary step as hominins learning to control fire. There is no turning back, and there never has been a case of a human society deciding to turn off the power (other than some small cults). There are still well over a billion people on Earth with intermittent or no electricity, but eventually they too will be electrified, because they want to be.

Can EMR in the atmosphere ultimately affect our genomes? In *The Developing Genome*, David S. Moore says "genetic, epigenetic and environmental factors... *operate as an integrated system*" (his italics), which includes "the cultures and physical environments in which we live out our lives,...i.e., the environments outside our developing bodies as well as the local [internal] environments surrounding our

DNA." The external environment can in effect penetrate our bodies by affecting our nervous systems and hormones, which in turn can cause epigenetic impacts on gene expression.

Moore cites some leading geneticists to support this point of view. "The susceptibility of the genome to epigenetic modifications provides a layer of genetic regulation that is sensitive to a lifetime of experiential and environmental factors." (Tania Roth). "Epigenetic studies make clear that the environment penetrates the genome at its core, and influences the expression or nonexpression of genes." (Marinus Ijzendoorn).

Many contemporary scientists have concluded that genes alone do not determine the traits we exhibit at birth and as we develop over our lifetimes. "Genes are not the blueprint for life," said Denis Noble, renowned emeritus Professor of Physiology and Biology at Oxford, in an influential article in the February 2024 issue of *Nature*. He says our biological systems influence our genes. Howard Bloom wrote in *Global Brain* (John Wiley & Sons, 2000): "Social experience literally shapes critical details of brain physiology, sculpting an infant's brain to fit the culture into which the child is born."

Genes by themselves do not control human biology. It is the sigmasphere interacting with our genes which together guide our development. The widespread adoption of electricity is driving heredity and evolution in ways we cannot ignore and scientists should explore. Our genomes may be unchanged from those of *Homo sapiens* who lived before the advent of the Electric Age, at least as far as we now know. But culture, the environment and epigenetics tell a more nuanced and compelling story: we have changed, we are changing, and evolving into a new species, *Homo electric*.

BACK ON THE MOUNTAIN

When we left our homely math professor on the mountaintop with the Wise Old Man, he wanted to know what his children would be like if he married his beautiful but not-so-bright girlfriend. Would they have his brains and her looks, or her brains and his looks? The Wise Old Man said, "It depends." The professor demanded to know, "Depends on what?" The Wise Old Man replied, "On everything."

The professor was totally frustrated by this answer. So the Wise Old Man told him that heredity wasn't what the professor thought, in his traditional gene-driven view about how traits, like intelligence and beauty, were passed on. He explained how the genome, epigenome, culture and the environment -- "everything" -- work together as a system to determine our traits at birth and as they continue to develop during our lifetimes.

The idea of hereditary operating as a complex system got the professor excited. "Maybe I can write a computer program to take all those factors into account and produce the answer I'm looking for about my children." But the Wise Old Man said, "My son, my advice to you is this. Forget about trying to figure out your heredity. If you love your girlfriend then marry her, and love your children too. Raise them in a loving home and a loving culture, and that more than anything will determine what they become. All you need is love."

The professor began to smile as enlightenment dawned on him. He thanked the Wise Old Man profusely. Then with a happy look on his face, and humming an old familiar tune, he headed back down the mountain.

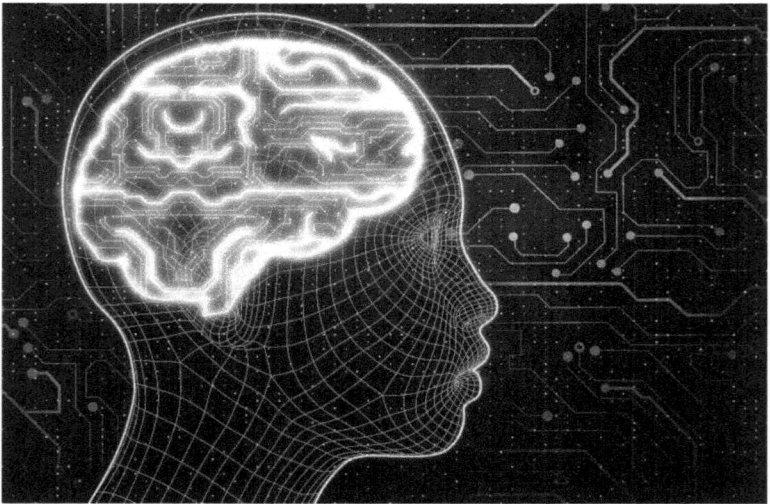

PART SIX

Homo Electric

Electricity is going to change everything. Everything!

-- *Tom Stoppard, playwright*

Chimpanzee

A DYNAMIC NEW SPECIES

Darwin's theory of evolution explained how natural selection and heredity lead to the emergence of new species, usually over vast stretches of time. The evolutionary lineages of human beings and chimpanzees, our closest primate relatives, diverged about seven million years ago, and it took about four and a half million more years until the genus *Homo* emerged. Since then, several species of *Homo* have come and gone, with *sapiens* the sole survivor for about 40,000 years, until its successor *electric* began to emerge.

Since we parted ways evolutionarily, chimpanzees have not experienced development nearly as far-reaching as that of humans. The success of a species can be due not only to the changes brought about through natural selection, but also to stability, an ongoing ability to thrive in a natural environment to which it is well suited, like the chimp has. Their species *Pan troglodytes* is still with us after seven million years. In that time there have been at least 21 species of hominins.

Many species of animals and plants have gone extinct due to major changes in the environment, like the onset of an ice age. Today, many are on the brink of extinction due to climate change and loss of habitat. They do not have the capacity to adapt. But *Homo electric* is uniquely and dynamically adaptable as a species. It is said that the only constant in our lives is change. We are not only subject to outside forces, as all living things are, but are active agents for our own change through technological and cultural innovation. We are altering our lives, our culture and even the course of our evolution. It is an electrifying development.

The Invisible Hand

116

Unnatural Selection

Natural selection is the driving force of evolution. It takes the variability of traits in any given species and favors the advantageous ones by passing them down to succeeding generations through genetic heredity, the "survival of the fittest." Whether we're talking about beetles, blowfish, bullfrogs, bobolinks or baboons, natural selection has caused all species to evolve to what they are, including the line of primates which became bipedal hominins and eventually *Homo sapiens*. But for *Homo electric*, a species not living in a state of nature, natural selection is being supplemented and at times supplanted by the "unnatural selection" of manmade choices.

Sometimes it seems as though natural selection is viewed by geneticists as an almost mystical force, an invisible hand that points to traits declaring "yes, that one" but "no, not that one." The idea of an invisible hand in the realm of economics was posited by Adam Smith in *The Wealth of Nations* (1776). He believed that if everyone pursued their economic self-interest, then competition would, through the invisible hand of a free market, result in economic prosperity. This was still the leading theory explaining economic behavior when Darwin developed the concept of natural selection. We have to assume that as a university graduate, he was familiar with it.

Smith formulated the principle of "absolute advantage," the ability to produce a good or service more efficiently than one's competitors, and as a result prevail in the marketplace. Is this not like the "marketplace" of evolution in which traits with an advantage for survival are selected by the invisible hand of nature, leading to the "prosperity" of a species?

Today there is no such thing as Smith's free market of pure competition, if there ever really was one. Governments play a major role in how economies function through laws and regulations, fiscal and monetary policy, and the influence of politics. So too do non-governmental organizations such as labor unions, business trade associations and nonprofits, as well as public opinion (where it can be expressed). These actors may constrain the invisible hand of self-interest for the presumed economic benefit of society, for the well-being of the species.

To Darwin, natural selection was a matter of life and death. He observed that because species tend to over-populate, those individuals with advantageous traits were more likely to survive and pass them on to their offspring. But unlike organisms living in a state of nature, human populations are subject to many unnatural factors affecting survival. Thanks to modern medicine and technology, 98% of children born in western societies live to childbearing age, a survival rate far greater than anything known in nature. They may well then pass on disadvantageous traits to their offspring, because we have interrupted natural selection with the human value that every being has an equal right to live (an ideal unfortunately too often unrealized). Unequal availability of medical care can result in the survival of the richest, as when the wealthiest nations secured the early supplies of COVID-19 vaccines.

I do not mean to imply that natural selection is kaput. But the process is complicated by "unnatural" human values which interfere with nature acting on its own. Darwin himself noted other forms of selection, like *artificial* selection, the cross-breeding of animals and plants to improve the species, and *sexual* selection, the choosing of mates, often in competition with others, through mating behavior and displays. *Social* selection is a transactional process, by which an organism offers something of value to another organism in exchange for the opportunity to reproduce. Has our professor been giving gifts to his girlfriend in an effort to win her over?

What we can think of as *cultural* selection is deliberate human intervention in the selection process, as with medical care, in accordance with the values of a culture. The eugenics movement in the early 20th century was a misguided attempt to "improve" the human species by encouraging procreation among some people while discouraging it among others, even to the point of forced sterilizations. Eugenics took its most extreme form with the Nazis, who selected the Aryan "race" by eradicating Jews and others they deemed inferior. Many of the victims no doubt had desirable traits which would have been propagated had nature been allowed to take its course.

There are many ways we choose to give people the chance to survive and reproduce, despite what nature might otherwise have intended. We have invested huge resources to develop and transplant replacement body parts. Traits which prevent reproduction, like sterility, constitute natural *deselection*, but they may be overcome by assistive reproductive techniques such as *in vitro* fertilization, involved in 2% of the births in the United States. CRISPR technology for editing DNA is a genetic revolution of no doubt far-reaching, but unknown, consequences. Can we speak of "natural" selection when we are capable of manipulating the functioning of our genes?

Culture will determine how this capability is used, the culture of the Electric Age. Like anthropologists, let's look at some characteristics of that culture and its people.

THE POWER OF LIGHT

We can all think of examples of how medical technology prolonged lives, regardless of their fitness to survive and pass on advantageous traits, bypassing the process of natural selection. One is of particular interest here because it involves electricity and its availability, or lack of, at the very moment when a new set of genes enters the world: childbirth. In 2008, some areas of Nigeria had power only about half the time, and there was a high rate of mothers dying in hospitals while giving birth. An American couple, Dr. Laura Stachel and Dr. Hal Aronson, developed a small solar electric generating system for hospitals there. Maternity deaths dropped by 70%.

That led them to produce a product called the Solar Suitcase, a portable solar electric unit to power lights, medical devices and mobile communications. Their foundation, We Care Solar, made thousands of Solar Suitcases available to serve health care facilities in over 50 countries. Creating this product, and financing and operating the foundation, is just one instance of conscious and collective social action to save the lives of all mothers and babies, without regard for their genetic makeup. The foundation selected them, not nature.

Photovoltaic systems, using solar cells which generate electricity when struck by light, are being increasingly deployed to provide power to buildings and to the electric grid, as well as in places where electricity is not reliably available, or not at all. We can thank Albert Einstein for this technology. He is, of course, best known for his General and Special Theories of Relativity, and you might well think that would have earned him the Nobel Prize in Physics. But the only

Nobel he did get was in 1921 "for his services to Theoretical Physics, and especially for his discovery of the law of the photoelectric effect," the physics of how light can produce electricity, the basis of photovoltaics.

There is an almost magical quality to this process. Generating electricity usually involves motion and sound, whether a dynamo turning, water flowing or wind turbine blades spinning. But with photovoltaics, photons of light move silently and imperceptibly, except perhaps for the sunlight we might see reflecting off the PV cells.

My first experience with photovoltaics was in the late 1970s, working for a nonprofit organization in Cambridge, Mass., called the Northeast Solar Energy Center (NESEC). It was a pioneering public-private sector partnership established with federal funding in the aftermath of the Arab oil embargo of 1973. Its mission was to increase awareness of renewable energy technologies and to catalyze their adoption. The president of the company which ran the center was an MIT-trained aeronautical engineer and entrepreneur, Lawrence Levy. During the Kennedy Administration, he was appointed by the president as minister to NATO, an ambassadorial-level position which reported directly to the Secretary of Defense.

As a member of the military-industrial-university complex, Levy recognized the security imperative for the United States to develop and commercialize alternate forms of energy to reduce its dependence on fossil fuels. He was instrumental in getting the funding for NESEC through his relationship with the Speaker of the House, Thomas P. "Tip" O'Neill Jr., whose Congressional district included Cambridge. O'Neill brought Levy's proposal to President Jimmy Carter, a solar supporter.

The author (left) at NESEC circa 1979 explaining the design of an experimental photovoltaic system to Denis Hayes, founder of Earth Day. The metal cones were used to reflect and concentrate sunlight on PV cells, so as to use fewer cells at a time when they were prohibitively expensive for most applications. The system was never deployed because cell prices declined.

I, on the other hand, was a long-haired member of the counterculture and former rock concert promoter, although a graduate of Harvard Law School. I came to solar from an ecological rather than a national security point of view, but Larry and I agreed about the importance of renewable energy. He hired me first to help write the grant proposal for NESEC, and then to serve as its Manager of Government & Public Affairs. Despite the "Solar" in our name, we worked with the whole spectrum of renewable energy technologies.

A lot of what we did involved applications which were already commercially practical, like solar water heating. Photovoltaics wasn't yet financially feasible, but I was fascinated by its promise, that you could install solar panels and generate electricity in a remote location with no electrical infrastructure, like say in the Andes. Working at NESEC I didn't have an opportunity to bring PV to Peru, but I did participate in our demonstration of the technology in the Adirondacks.

This was during the 1980 Winter Olympic Games in Lake Placid, New York. PV panels supplied power for race timers during the women's giant slalom, the first-ever use of solar during an Olympics. The purpose of the demo was to create awareness of the technology by seeing it at work in a highly-visible application. I was already a believer.

By the 2010s, the cost of generating electricity by photovoltaics was decreasing dramatically. PV was not only viable on a large scale for use by businesses and utilities, but the installation of small-scale residential systems became the fastest growing industry in the U.S. Over five million households now produce their own electricity with solar. We've come a long way from making fires in our caves.

Woven image of Joseph Marie Jacquard produced by his eponymous loom "programmed" with 24,000 punched cards

THE COMPUTER GENUS

No, that's not a typo. I do mean "genus" in the taxonomic sense, not "genius," although there certainly was lots of that involved in the development of computers. Earlier I wrote about cultural evolution as part of the sigmasphere affecting gene expression. Technology is a major part of our culture, and no technology has been more important to the evolution of culture and humans in the Electric Age than computing. So let's consider the evolution of computers, much like we did the evolution of the genus *Homo*.

Just as there have been several species within *Homo*, so too have there been several species within the genus *Computer*. Like some species of *Homo*, they have evolved along separate branching tracks (think *neanderthalensis* and *sapiens)*, sharing certain traits.

Like *Homo*, the genus *Computer* had ancient predecessors. In 1822 the English polymath Charles Babbage conceived of a mechanical (non-electric) computer, the "Difference Engine," and then in 1837 a more advanced "Analytical Engine." Some of the latter's "traits," such as programmability and integrated memory, are still essential characteristics of computers today. Neither machine became fully operational during Babbage's lifetime, but some "bones" from his prototypes are preserved in the Science Museum in London. Like reconstructing a pre-human "Lucy," two complete models have been built from Babbage's designs which prove that they would have worked.

An innovation incorporated in the Analytical Engine was programming it with punched cards, which were first used to control the loom patented in 1804 by Joseph Marie Jacquard.

In 1955, these 62,500 punch cards contained 4.5 megabytes of data. Today for about $100 you can buy and hold in your hand a small hard drive storing a million times more data.

A young woman in London named Ada Lovelace became friends with Babbage and developed an algorithm for the machine to perform certain mathematical calculations, run by punched cards. She is usually credited as being the first computer programmer. She was not only a math whiz but a writer as well: "the Analytical Engine weaves algebraical patterns just as the Jacquard loom weaves flowers and leaves," as she put it. Ada likely came to her literary aptitude through genes inherited from her father, the poet Lord Byron.

More than a century later punch cards (their name now shortened without the "-ed") were still being used to enter data into computers. A young woman on Long Island, New York named Maureen Tucker earned her living in the 1960s as a punch card operator, developing rhythmic chops which served her well when she became the drummer for a new band called The Velvet Underground. Their music wove a hypnotic spell, and we'll revisit them later.

Early in the 20th century mechanical calculators began to incorporate electric motors. But the concept of the modern computer, run by electricity, was first theorized by the Brit Alan Turing in 1936. During World War II the need to crack the encoded radio telegrams of the German military command drove development of a series of machines called Colossus, considered the first electronic programmable digital computers. Put into service in England in early 1944, they were operated by women, and were essential to the war effort.

A much more powerful computer made its debut in early 1946 at the University of Pennsylvania. It was called ENIAC – Electronic Numerical Integrator and Computer – and it weighed 30 tons, contained over 18,000 vacuum tubes, and was dubbed a "Giant Brain" by the press. This was an early appearance of that species of the genus *Computer* which came to be known as *mainframes*, room-sized behemoths which stood astride the world of computers and eventually evolved into today's supercomputers.

Only governments and large institutions could afford to buy and operate these machines, the need for which was often driven by military applications. Nor did smaller institutions and businesses have any use for the massive computing power required to, for example, model thermonuclear explosions, as ENIAC indeed could. Then in the 1960s a new species arose: *minicomputers*. Refrigerator-sized, they dispensed with unreliable vacuum tubes which gobbled up electricity, adopting new technologies such as transistors and integrated circuits. These advantageous traits enabled them to achieve dramatic reductions in cost and size, and to thrive in the marketplace. They could model financial explosions of start-up companies.

But computers were still the tools of organizations, not individuals. In 1960 I took a college computer course. You wrote a short program with pencil and paper, and it was then sent to the computer lab, a mysterious place you were not allowed into. There the program was entered into punch cards and run overnight. Two days later at the next class you might learn that due to a single incorrect character, your program didn't run. Try again next week.

Our relationships with computers would become far more immediate, interactive, satisfying and personal. Evolution is a slow process, yet sometimes it can run so fast that it needs to be spelled with an "R" in front.

The Microcomputer Revolution

Just before Christmas in 1974, the new issue of *Popular Electronics* appeared on newsstands and arrived in mailboxes. The cover story announced: "Project Breakthrough! World's First Minicomputer Kit to Rival Commercial Models." In fact, the subject of the story wasn't a minicomputer at all, but an even smaller machine called a *microcomputer*, a name which was not yet widely in use. But it certainly was a breakthrough, the emergence of a new species of the genus *Computer*. It was a Yuletide gift to the world which became popularly known as the personal computer.

The Altair 8800 pictured on the magazine's cover became the first commercially-successful microcomputer, a term referring to its single integrated microprocessor circuit (chip) as its central processing unit (CPU), rather than the more complex multiple processors which ran larger-scale computers. Not that the device was for everyone. It was sold as a kit that had to be assembled, a product for serious hobbyists.

The Altair had no keyboard or monitor, so here's how you entered data. Every letter or number can be represented by a series of eight bits, that is, a specific sequence of eight 0's and 1's. To enter a letter, you'd set eight toggle switches to "0" or "1" to represent the sequence for that letter, and then move another toggle switch to enter it. Okay, are you ready to input your next letter?

Despite these limitations, the Altair did have some noteworthy features which made it a true ancestor of the personal computers which followed, like an Intel CPU chip. An 8" disk

The author's Altair 8800 computer with 8" floppy disk drive

drive was available for it, the size floppy in use at the time. Its internal communication system, known as the S-100 bus, became for many years the standard for personal computers. The 8800 could be programmed with a piece of software called Altair Basic, developed by a Harvard student, Bill Gates, and his good friend Paul Allen, after they read the *Popular Electronics* story. It was the first product from their new company, which they named Micro-Soft.

Two young guys on the other coast, Steve Jobs and Steve Wozniak, had a vision for an evolution of the microcomputer that would be much easier to use. Their Apple II came pre-assembled, with a keyboard, a monitor which could display color graphics, and audio. It was truly a device you could take home, plug in and use. Launched in 1977, sales of the Apple II and its peripherals, accessories and software grew from $775,000 in its first month to $118 million a year in only three years. Its two competitors were also successful: the Commodore PET (Personal Electronic Transactor) and the TRS-80 from Radio Shack (fondly known to geeks as the "Trash-80"). The three machines proved beyond any doubt that, despite corporate IT guys sneering at such dinky little devices, the personal computer was more than fit to survive.

The PC business was booming, yet IBM was skeptical about the need for personal computers; it was hugely profitable from leasing large machines and software to businesses. But in response to their customers' demand for an IBM product, they decided to make one, although they projected that the total market for it was only 25,000 units. The IBM Personal Computer, introduced in August 1981, was unlike anything the company had ever made before. It sold in retail stores. Its open architecture, unlike IBM's usual proprietary designs, turned out to be the basis for its success, allowing third-party developers to create software for it. The IBM PC quickly became the alpha male of microcomputers. Many clones of the PC arose, hoping to bite off some market share.

In November 1983, I was in Las Vegas attending the computer expo Comdex, along with 125,000 other people. Excitement was in the air; the rate of innovation and growth in the PC business, as seen on the exhibit floor, was unprecedented. I ran into someone that I knew from many years ago, Andrew Singer, who was my younger brother Peter's best friend in their high school days. As we chatted among the throngs rushing around, Andrew was interested to learn that I was involved with marketing and public relations.

He told me he could really use my help on a project he was working on, although he didn't give me any details about what it involved. I explained that I was deep into developing a business plan as a consultant to Pat McGovern, the founder of Computerworld and many other tech magazines. I would be working on the project straight through the holidays, and couldn't think about taking on any more work at that time.

But Andrew was very persistent and wouldn't be put off. Since his office was only about ten minutes from my house down Route 128, the technology backbone of the Boston area, I agreed to skip lunch one day and stop by. When I arrived at the offices of his company, THINK Technologies, he sat me down in a conference room, had me sign a Nondisclosure Agreement, and left. He returned a few minutes later pushing a cart with something on it covered by a cloth. I was hoping there was a sandwich under it. Removing the cloth with a dramatic flourish, he asked me, "What do you think this is?"

I was puzzled. It was a boxy object, in a plastic case about a foot high. It had a small built-in screen and a keyboard, but didn't look like any computer I'd ever seen. And it had a palm-sized object attached to it by a wire. Before I could say anything, Andrew answered his own question. "This is the new computer from Apple. It's called Macintosh."

It was to be introduced on January 24th at the company's annual shareholders meeting. THINK had been working under contract to Apple to develop programming tools for the Mac (versions of Pascal and C) to enable software developers to create applications for it. Without them, the machine was nearly useless. Andrew explained its graphical user interface and the function of that little object, called a mouse. He asked me to put together a press kit for THINK, since the meeting would attract a lot of media coverage, and to accompany him to the event in California. I didn't hesitate to say yes.

Cupertino, CA, January 24, 1984. Andrew and I are sitting in a large auditorium packed with Apple shareholders, employees, press, industry analysts and computer executives. Among the special guests are the heads of four leading software companies who have developed products for the Mac, including Bill Gates of Microsoft, plus a fifth company, THINK. The sense of anticipation is electric. Two days before, during the Super Bowl, Apple had aired a dramatic "1984"-themed 60-second spot announcing the launch of the Macintosh, but the ad didn't show the machine itself.

Steve Jobs comes on stage and after a few brief remarks goes over to a carrying case propped up on a platform and lifts out a Mac. Other than members of the small development team, no one in the audience has seen one before. They clap. Jobs holds up a disk, smaller than the 5¼" floppies we're used to using with PCs. More applause. He then inserts it into the Mac drive to run a graphical demo of its capabilities, with a musical sound-track, all generated by the Mac and projected on a large screen. When MACINTOSH in bold caps scrolls across the screen, the audience claps even louder and hollers.

Near the end of the demo, the Mac speaks in a computer-generated voice and, after cracking a couple of jokes, concludes by saying, "It is with considerable pride that I introduce a man who's been like a father to me, Steve Jobs." The audience leaps to their feet, clapping and cheering wildly, for Jobs and Macintosh. It was like the conclusion of a rock concert. I half expected people to hold up lighters.

Mutations usually occur unexpectedly and even unnoticed. But the Macintosh mutation of the *microcomputer* arrived on a preordained date to great fanfare. Everyone in the auditorium knew they were seeing the future of personal computing. Macintosh embodied Adam Smith's "absolute advantage," and it would not be long before the entire PC market would select for the traits of the Mac: graphical user interfaces, mice, windows, icons, fonts. Easy to use, but powerful.

Back at the ranch, THINK engineers were herding code for the first e-mail package for the Mac. E-mail was available then only in organizations with large computers, through desktop terminals. In most offices, electric typewriters and photocopiers were the only significant innovations in decades. Intra-office communications were typed up, photocopied and distributed by mail carts with other papers like letters and magazines. That inbox on your desk was the nexus, the place

The author's Macintosh 128k used while working with THINK

where all the paper arrived, as well as a makeshift task manager holding documents you'd get to later.

THINK's e-mail product would eliminate much paper communications. It ran on the Ethernet local area networks which were starting to become standard in many offices. We named the new product "InBox" and positioned it as "desktop communications software," a play on "desktop publishing," for which the Mac was renowned. E-mail reduced the use of that longtime office mainstay, the telephone. No longer did you have to get on the line voice-to-voice with someone in the office, playing telephone tag to connect. Just send an e-mail to their InBox, or many InBoxes at once, and await their reply.

Soon e-mail could be exchanged with people outside the office. That's when someone realized that they could send millions of messages at virtually no cost. In simpler times we consumed a food called Spam. Now we were consumed by it.

POCKET THE DIFFERENCE

Andrew Singer was a serial innovator. Mac Pascal and its successor THINK Pascal became the leading development environment for Macintosh, in large part due to its graphical user interface, for which Andrew was lead designer. A GUI was new to programming languages yet so appropriate for the Mac and enabled a thousand applications to bloom. E-mail became a standard application for Macintosh and every other personal computer. After THINK was acquired by software powerhouse Symantec, he worked for Radius, a computer hardware company founded by several members of the original Mac development team. There he invented the Pivot, a monitor which could be physically tilted from landscape to portrait position, with the image on the screen automatically reorienting in real time, along with the menus and the mouse.

Steve Jobs, of course, was the greatest innovator in the evolution of microcomputers. With Steve Wozniak he created the Apple II, one of the first mass-produced products which put computing power in the hands of consumers. Then he led development of Macintosh, which transformed personal computing with its mouse-driven graphical user interface, and which, with its LaserWriter printer, introduced desktop publishing.

Good design was always a trait of the products Jobs had produced at Apple, both in their physical appearance and their functionality. After leaving the company in 1985, he funded the spinoff of Pixar from the computer graphics division of Lucasfilm. Pixar took movie visual effects to another level with "Toy Story," the first 3D computer-animated full-length feature film, nominated for three Academy Awards and the recipient of a Special Achievement Award from the Academy.

Jobs returned to Apple as CEO in 1997, when it was on the brink of failing. A year later he began its turnaround with the rebirth of Macintosh as the iMac. With iTunes, launched in 2001 and soon followed by the iTunes Store, he brought digital music to personal computers, while shaking the foundations of the traditional music business and its reliance on physical media like CDs. The little portable iPod took music off the desk and into the street, making it laughable to remember old music players like the Sony Walkman and those giant boomboxes people lugged around.

At the launch of Macintosh in 1984, Jobs spoke of the telephone as the first desktop appliance, and his dream of making Macintosh the second. Well, if you include all the personal computers which adapted to the Mac mutation, he succeeded wildly. Then in 2007 he took those two products and squeezed them into a handheld device called the iPhone.

There had been earlier versions of "smart" phones, like the Blackberry, designed primarily for business customers, with a physical keyboard and small screen. But just as the Mac was to personal computers, so too was the iPhone to cell phones: a transformative mutation. Incorporating technology from the Pivot monitor, it could switch from portrait to landscape display in real time.

For the purposes of our little exercise in computer evolution, it's worth asking, I think, whether the iPhone is a mobile phone with added features, or a computer that can make phone calls? I'd classify it as a new species of the genus *Computer*, designated the *smartphone*. It can do just about everything a PC can do and more, like take photos. Just as dinosaurs and massive mammals made way for smaller creatures better adapted to survive, so too did Colossus and ENIAC make way for handheld devices like the iPhone, more powerful despite their much smaller footprints. You can pocket the difference.

There is one legendary yet often forgotten player in the microcomputer revolution that deserves mention here, a research and development division of Xerox established in 1969 and called the Palo Alto Research Center (PARC). Its mission was to create "the office of the future," and many wondrous things emerged from its labs: a personal computer with a graphical user interface, a mouse to control it, a laser printer, and the Ethernet local area network.

For whatever inexplicable reason, Xerox never did anything with these inventions, and by the time the company tried to enter the personal computer business many years later, they had missed the window of opportunity. In 1979 Steve Jobs went to a demo at PARC and recognized the value in what he saw there, which in a few years became the heart of the Mac. Some people denigrate Jobs as having "borrowed" this technology from PARC. But he had the vision to put it all together in a real-world product which had enormous success in the market and changed the nature of personal computers.

Some archaic human ancestors saw what others didn't in a stone, the possibility to make it into a tool, the Oldowan hand axe. Other humans subsequently refined that technique to further shape stone into Acheulean tools. For Jobs, there were many stones that became the Apple II, Macintosh, iPod and so on, as we've seen. They helped evolve *electric* culture.

Cultural evolution had earlier developed the skill of reading, leading to changes in the brain to support that activity. This involves "a culturally constructed network," the brain's "letterbox," which is located in the "visual area" between "a region specialized to recognize faces and one focused on objects." (Henrich, *The Secret of Our Success*).

Technological evolution has produced the personal computer and the smartphone, both of which require facility in reading and in recognizing symbols like icons. Graphical user interfaces with controllers like mice and touchscreens better utilize a wider range of the brain's visual capabilities than earlier interfaces based solely on alphanumeric characters, commands and keystrokes.

We've become attached to such brain-friendlier computers and smartphones. They feel good to use, stimulating our brains like a laboratory mouse pushing a lever to get a food pellet. Henrich says that changes in the brain from reading are biological, not genetic. But a *culture* of reading is passed down to succeeding generations. Electronic devices also cause changes to the brain, even more so than reading. Whether biologically or genetically, or perhaps epigenetically, they will ultimately affect gene expression and, thereby, evolution. Children seem born to using these devices. They are a defining trait of the culture of *electrics*.

We have evolved from our ancestors of 200 years ago, just before the Electric Age began in about 100 BC, one hundred years Before Computers. A blink in the course of human history. Yet we can hardly recognize ourselves in those "archaic" humans. They would not easily comprehend us. Our world would seem magical to them, the magic of electricity.

Ruins of an abandoned 20th century office

REMOTING

For many people, the coronavirus pandemic caused a paradigm shift in working and schooling, from on-site to remotely from home, made possible by personal computers and smartphones with fast connections. Remoting is not a passing phenomenon which will fade away with the virus. It is a characteristic of *Homo electric*, the first species of humans who interact regularly in real time without being face-to-face.

After our ancestors began living in settlements about 12,000 years ago, working at home was the way of life for farmers, and then independent craftspeople and merchants living above the store. As the Industrial Revolution got going in the mid-to-late 18th century, people left home to work. This trend accelerated in the 20th century with the growth of cities and suburbs, and workers commuting by car, train and bus. The coronavirus slammed the brakes on that, if only for a short time. "Rush hour" and "traffic jam" no longer aptly described roads now uncongested at the height of the pandemic. The air took a deep breath. We got a sense of what is possible. Now we need to learn how to make remoting better.

In recent years, even with technology like e-mail which made working at home more feasible than ever, most employers were resistant to the idea. They were used to being able to keep an eye on their underlings, face-to-face if need be. You'd almost think there's a gene in bosses' DNA compelling them to keep their workers close at hand. But suddenly they were confronted with a mutation in the culture of work, caused by a virus. Survival for office workers meant staying home and remoting. Spouses and kids were there too.

The work went on, even from laptops perched on kitchen tables. Many employers were surprised to find that employees working remotely were often more productive than they had been back at the office. They took notice, and some began rethinking the concept of an office, motivated by potential cost savings from renting less space to accommodate staff on-site only part-time or occasionally. For enlightened employers, increased employee satisfaction was a plus.

The evolution of the office has in fact been going on for centuries. The word "office" was first used by Geoffrey Chaucer in *The Canterbury Tales* in 1395, to indicate a place where people conducted business. The first "modern" structure purposely built to house offices was erected in 1726 in London. The classic office arrangement entailed a large space filled with employees' desks, neatly lined up like the desks in a school room. In the 1960s that mutated to a room filled with cubicles. Subsequent innovations like the open plan and co-working spaces followed.

Now office designers are faced with a new challenge, an office some of whose employees work from home some of the time. Remoting is taking root in the work world, as it yields benefits to employers and employees alike. Some major tech companies told employees they could continue working from home even after the pandemic, although we've seen bosses reneging on that offer. Video interconnection apps like Zoom and Teams are proving to be workable replacements for many of the face-to-face meetings in the office.

Some people lament that with remoting we lose those chance encounters in the hallway which serendipitously lead to great new ideas. But I think the role of serendipity at work is exaggerated. There are examples to be sure; Bell Labs is a classic one, but that's an unusual environment of researchers and innovators. Casual chats around the typical office are likely to be about TV shows, sports, social media, gossip. Beyond formal Zoom sessions, *electrics* will use electronic communications informally in ways that foster serendipity. People will still gather in person at the office to work together, exchanging ideas and building personal relationships face-to-face. To be continued later online.

Parents working at home found their kids schooling at home. Educational institutions at all levels were among the first organizations to shut down their physical facilities. But school was not out. Students were required to continue their education by remote learning. Results with this approach have been very mixed. Some students did well on their own at home, many others did not. They had no prior training in how to learn remotely, they were simply told to go home and do it. Some students had computers, others did not. Some had fast internet connections, others had slow ones or none at all.

Such disparities among K-12 students' abilities and facilities, as well as the need for kids to socialize, led to pressure in many school districts to reopen schools. To do so should have been primarily an issue of health and safety, but it got tangled up in politics. Some districts ping-ponged between opening and closing, with hybrid part-time solutions in the mix. Caught in the middle were the teachers, who had no training in remote teaching. They might have done some interesting educational lessons, like interconnecting kids from far-flung areas. But nobody fully explored the potential of remoting, because nobody was vested in its success.

That some students are ill-equipped, skill-wise and technology-wise, to be effective remote learners is an understandable reason in the short run to send them back into a school building. For young children especially, socializing in school is an important part of their development. But as they go through high school and college, they will need to become better remote learners, a skill schools should teach, because down the road they'll be participating in a workforce where remote interactions are common practice.

To enter that workforce, it has been the case for some time now that job candidates are often interviewed remotely by video, not face-to-face. Someone who excels in this remote process, it might well be assumed, will likely be an effective participant in the online conferences which are proliferating to replace in-person business meetings.

To facilitate remote learning, every student will need a computer. And bit by bit, upgraded internet infrastructure is being deployed to bring faster and more reliable internet service to everyone everywhere. I know the problem all too well, living in a rural area and being one of the founders of WiredWest, the municipal cooperative I mentioned earlier whose mission was to deploy fiber-optic broadband networks in the small towns of western Massachusetts. I finally got connected on July 14, 2020, Bastille Day, freed at last from the prison of slow and unreliable internet.

Remote learning and work were necessitated by the unique circumstances of the pandemic, but remoting will increasingly be how *Homo electrics* participate in the economy and society. We saw with those uncongested roads an opportunity to reduce the use of fossil fuels. It will become a necessity.

Bringing Education to People

Remote (or "distance") learning did not appear out of nowhere with the pandemic; it had a history. I had the good fortune to know a true pioneer in the field, Glenn R. Jones, one of a group of cable TV entrepreneurs in Colorado called "the cable cowboys." In 1967 he bought a small failing cable system. To come up with the $1000 down payment, he had to take out a $400 loan on his Volkswagen. Twenty-five years later his company, Jones Intercable, served 1.4 million customers, then the ninth largest cable operator in America.

Glenn called the market he served "the human mind," the brain "a three-pound wet computer." He was fond of saying, "Don't bring people to education, bring education to people." In 1987 he launched Mind Extension University, which aired college courses for credit on cable TV. It was only available as a one-way telecast, a limitation of the technology available then. But in 1993 he founded Jones International University, the first internet-only accredited degree-granting institution of higher learning. A computer was your classroom.

He later played a major role with the Library of Congress in creating the National Digital Library, which digitized its collections to facilitate onsite and remote access. For his vision and many achievements in distance learning and telecommunications, in 2015 Glenn was honored as a "Library of Congress Living Legend." He died a month later. Jones International University also passed into history that year, but many of its programs were transferred to the online institution Trident, which named one of its divisions the Glenn R. Jones College of Business. A fitting memorial.

PARANOIA

In the last few years, a particularly pernicious form of what was once just rumor-mongering has become a dangerous phenomenon: so-called "conspiracy theories" spread on social media, like Facebook and X (formerly Twitter). This new way to interact was gaining traction with users when the iPhone was launched in January 2007. While people can and do use social networks on their personal computers, the rapid adoption of smartphones drove their usage exponentially, as well as that of the many other social media services which soon arose and attracted huge user bases.

To call an unsubstantiated rumor posted on social media a "conspiracy theory" is a misnomer, a term which gives it a credibility it does not deserve. It is by no means a "theory," because a theory has a testable hypothesis. It is better termed a "conspiracy fantasy" or, all too often, a "paranoid delusion," like the infamous pedophile ring supposedly operated by Hillary Clinton out of the basement of a pizza parlor in Washington, DC. The danger to society is that in a delusional state, people may react uncontrollably to what they perceive as a threat, tilting at windmills, or worse, shooting at them.

Paranoia has a long history in human evolution, and in the Electric Age can spread like, well, a virus. The following section is based on two papers on paranoia and digital media by Dr. Peter L. Nelson. He is a neuroscientist, social scientist and psychologist in Australia, and my brother.

The papers are available at socsci.biz/paper1.pdf and socsci.biz/paper2.pdf.

FDR's first Fireside Chat, March 12, 1933

FEAR ITSELF

By the early 1930s, more than half of all U.S. households owned a radio. To calm a nation deeply shaken by the Great Depression, shortly after taking office President Franklin D. Roosevelt initiated a series of radio addresses known as "Fireside Chats." They were the Electric Age equivalent of a tribe gathered around the campfire to hear from its chief. In rapt attention, American families huddled around their radio sets, prominent fixtures in their living rooms.

In his address at his inauguration, FDR famously said, "We have nothing to fear but fear itself." He was of course speaking to the crisis facing the country. But his words reflect a tendency since our earliest ancestors as to how we perceive and respond to danger, real or imagined. It is what Dr. Nelson refers to as the "paranoid tilt" in our cognitive processing. "Fear itself" is playing an increasing role in how many people now think and behave, especially in their interactions with social media.

Pre-humans and early humans were preyed upon by large carnivores. To understand how they might have reacted to this threat, Nelson suggests we look at prey animals today. Imagine that you are observing a herd of Sambar deer grazing in an open meadow in east India. This is a dangerous place for these herbivores, vulnerable in such unprotected ground to their most powerful predator, the Bengal tiger. The adult deer often stop and raise their heads, listening for sounds and sniffing for odors wafting through the air.

At the sound of a breaking twig, they have to decide instantly if it is just a monkey stepping on it or a tiger moving towards them, whether they should flee or return to grazing. Those deer slow to react will soon be culled from the herd by tigers and removed from the gene pool by natural selection. But some are too quick to react, rushing off at the slightest sound. By over-reacting too frequently they become fatigued, and by separating themselves from the herd lose the protection of numbers, thus leaving them vulnerable. Those deer who are "just fast enough" are best equipped to survive, and that characterizes most of the herd. The overly-quick reactors, however, will tend to survive better than the overly-slow reactors, leading to many more over-reactors than under-reactors in the general population. Over-reacting can be a means for survival.

Unlike deer, we humans do not just resume grazing after a dangerous incident. We construct a narrative about the episode, because we are story-tellers, it is how we perceive the world. We think about it, compare notes with others, and express a range of emotions. Dr. Nelson asserts that humans' deep evolutionary roots as prey for predators have bestowed on us an acute danger alert and response system. This can lead our highly-evolved brains to form paranoid ideas about what happened, or might happen next time.

By "paranoia" he refers to our capacity for creating beliefs and scenarios in which malevolence and threats are ascribed to people, organizations or other forces where there is no clear and verifiable danger, but rather a supposed hidden danger inferred from a pervasive sense that something threatening is operating behind the scenes. Typically, such beliefs are held to be true with an unshakeable certainty, no matter what evidence to the contrary may be offered.

Paradoxically, our higher neurological evolution, driven by ancient danger alert systems in our brains, results in a propensity for paranoia and over-reacting to perceived danger, a "paranoid tilt" in our thinking and the stories we tell ourselves and others. This is amplified when we interact with digital social media, in which reality-testing is difficult, if not impossible. The speed with which such interactions take place and the massive volume of incessant online postings gives us neither the time nor the ability to readily assess the validity of what we're being exposed to. Instead, it may trigger our alert systems, resulting in a positive feedback loop which Dr. Nelson calls "digital neural-amplification." In this way, conspiracy "theories" promulgated on the internet can be seen as paranoid self-stories run amok.

In our ongoing internal storytelling through which we view the world, there is a tendency for us to fill in more than what our senses have perceived, imagining, for example, a source for a threat when supporting facts are unclear or missing. As we evolved from living in small bands of hunter-gatherers to participating in large-scale highly-complex societies, danger signals became more difficult for our ancient warning systems to interpret, and our self-stories took on meanings beyond what was actually occurring. This ambiguous situation fed the paranoid tilt in our cognitive processing, as "fear itself" came to play an increasing role in the ideas we formed and the stories we told and believed.

Like the vast majority of the deer whose reactions are "just fast enough," Dr. Nelson notes that for most of us the reactivity of our nervous system is finely balanced, and we can manage the signals from our alert system by appropriately assessing the danger. Inhibitory neurological mechanisms moderate our alarm responses, preventing an out-of-control feedback loop in which paranoia takes over some people and dominates how they perceive the world. This extreme over-reaction found support on social media from the likes of the anonymous poster "Q." Identifying themselves as QAnon, followers of this "conspiracy theory" attributed their sense of threat to hidden forces of the "deep state" controlling the government, and their lives.

A system like Facebook is designed to encourage user interaction and amplification of their responses to what they see. They may be drawn in by "click bait" (such an appropriate term for humans as prey!), which could be an alarming news item to which they respond by "liking" it. Then they are fed more such stories which in turn they "like," and so on, in a deepening spiral driven by Facebook algorithms down the rabbit hole into a paranoid digital Wonderland in which Q the Queen of Hearts dictates what is real or not.

Social media users like QAnons talk about "doing their research." But in reality what they do is to selectively find self-reinforcing but manufactured "facts" to prove the "truth" of what they fear. "Children are being held hostage in the basement of a pizza parlor!" They know it must be so, because they read it online. But as one young man found when he showed up with a gun to rescue them, there were no children in the basement. Indeed, there was no basement, except in the depths of his paranoia.

Through the Cognitive Revolution or other evolutionary developments, *Homo sapiens* acquired many new mental capabilities, including the ability to recognize oneself as a unique individual, and to construct self-stories about who we are and how we relate to the world around us. Gossip became the way we shared and scrutinized those stories, our own and those of others. Like primates grooming each other, plucking insects from each other's fur, gossip was how we "groomed" and socialized each other. It was how we arrived at what we considered to be a shared truth, face to face.

Electricity disrupted that. It was not that we stopped gossiping, but that new sources of information were introduced into the conversation. Mass produced newspapers rolled off electric printing presses, with divergent points of view. Radio and then television added their voices. In the "Golden Age" of TV in the 1950s and 1960s, the three broadcast networks reflected very similar views. We all shared the central dogmas of our culture. But with cable TV and its proliferation of channels, and then with the internet and social media – where "alternate facts" abounded – the shared consensus was undermined, if not obliterated.

People now believe wildly different versions of "truth." FBI intercepts of communications among those who stormed the Capitol on January 6, 2020 showed fear that as president, Joe Biden "will destroy our way of life." Joe the tiger. Fear itself manifest. Many people who were not there that day share that fear. But aren't those who believe in such paranoid fantasies the real threat to our way of life?

Artificial intelligence is likely to worsen their paranoia. When an AI bot searches for "facts" to respond to questions, it certainly will encounter conspiracy theories. If it does not recognize them for what they are, then repeating them will give them credibility in the minds of the paranoid.

Bo Diddley

WE GOT RHYTHM

Music has been a part of every human society since the earliest days of *Homo sapiens*, and perhaps long before. Yet the role music plays in human evolution is not obvious. It even puzzled Charles Darwin. In *The Descent of Man* he wrote: "As neither the enjoyment nor the capacity of producing musical notes are faculties of the least use to man in reference to his daily habits of life, they must be ranked among the most mysterious with which he is endowed." In other words, just what does music have to do with the survival of the fittest?

Darwin concluded that for humans, music was all about sex, as with birds, who emit calls and sing songs in pursuit of a mate, as do many other creatures. That's an odd conclusion from someone who lived during the golden age of classical music. It's hard to imagine that the women in the audience at a concert were breathing heavily, ready to surrender themselves to members of the orchestra. Well, perhaps if the performer was the handsome Hungarian pianist Franz Liszt. He created such a frenzy among his female fans that the poet Heinrich Heine called it "Lisztomania" (more than a century before Beatlemania).

Other evolutionists looked at the role of music in tribal societies, where it was always accompanied by dance. They concluded that its primary function was to foster group solidarity, and in so doing to improve the survival chances of the group and its individual members, whether by cooperating in their daily lives or united in battle. If you've been to a rock concert, a rave or any other popular electric music event, especially where there is dancing, it's easy to see how it's about both sex and solidarity.

Where did music come from? Its origins have been much explored and debated, often linked to the origination of language. They do have certain characteristics in common as to how they're processed by the brain, especially after music incorporated lyrics. But I think music originated long before language, with a pre-*sapiens* species of *Homo*. Its genesis is in rhythm, not in words or melody.

Ancient humans lived in a world of rhythm: the beating of their hearts, mating, the sound and feel of their footsteps as they walked or ran (try doing it without rhythm!). Then *Homo habilis* and *erectus* became tool makers, chipping stones to form implements like hand axes, which they used for chopping and scraping.

Think about how you perform a repetitive physical task like chopping trees or carrots: it's always easier to do it rhythmically, your brain and muscles coordinating better to anticipate the timing of the next chop. So hominins not only experienced rhythm but produced rhythmic noises while making tools -- chip chip chip -- and using them -- chop chop chop.

Perhaps they made guttural sounds while working. *Erectus* had evolved to the point of being able to control the flow of air from the lungs to make vocal sounds loud enough to be heard. At some point in time some early humans banged stones together to make rhythmic noises for their own sake, grunting or chanting along. Then others started moving their bodies to the beat, and stomping their feet. A Stone Age rave.

Before going any further, I need to clarify just what I think music is, and isn't. I do not consider songs by birds and whales to be music. Those sounds are *musical* to be sure, but not music in the human sense. Human music has the potential for two or more people to make it together, to sing and play instruments in unison. That requires rhythm. We have rhythm but with very few exceptions, like parrots, animals don't. Woodpeckers are rhythmic for short bursts, for the same biological reasons we chop rhythmically, but woodpeckers don't synchronize their pecking in duets or trios. Chimpanzees are the animals most closely related to humans, but they don't have rhythm; experiments to train them to follow a beat all failed. Bluebirds can sing solo, but they can't do bluegrass harmonies. Music in its complete sense is an evolved human activity. We got rhythm, we got music.

It's hard to say when human singing arose, because it left no physical evidence. We do have a Neanderthal flute made from a cave bear bone 60,000 years ago. Before that they may have used hollow reeds or wood, but they were not preserved.

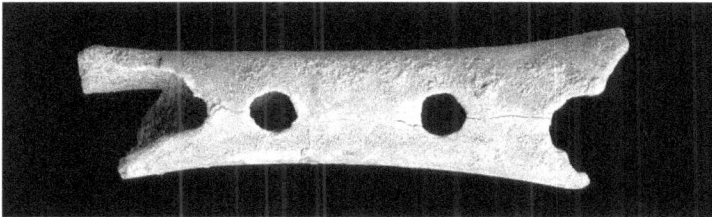

Flute-and-drum is likely the earliest form of music-making which arose after simple chanting. It can still be heard in the highlands of Peru, a traditional form of folk music in which a musician carries a large drum he bangs with one hand while playing a long wooden flute with the other. Hearing it echo in the hills around Vicos transported me to an ancient time.

Musicians in Vicos, with a couple dancing

ELECTRIC MUSIC

In Darwin's day, symphony orchestras played a wide range of artistically-crafted instruments, many still used today. Music was characterized by melody, harmony, tempo and dynamics. Once-prominent rhythm had largely slipped beneath the aural surface, still enabling musicians to play in unison but only occasionally dominant, like timpani in a symphony finale.

In the 20th century new forms of more rhythmic music began to emerge. The 1920s are often called the Jazz Age, the 1930s the Big Band or Swing Era. These were still acoustic forms of music, but their composers and performers were influenced by the more frenetic social energy of the Electric Age, and by the radio and records electricity made possible.

Acoustic guitars were often part of the lineup in jazz combos and big bands, but were hard to hear above the roar of horns, the pounding of pianos and the throbbing of drums. In the 1930s there were attempts to solve this problem by using primitive pickups, or by playing into a microphone. But in the 1940s two men – Leo, a tinkerer in California who couldn't play an instrument, and Les, a skilled guitar player in New York who couldn't play the sounds he heard in his head – set out to give a new voice to guitars with electricity. Their story is well told by Ian S. Port in *The Birth of Loud: Leo Fender, Les Paul, and the Guitar-Pioneering Rivalry That Shaped Rock 'n' Roll* (Scribner, 2019).

What resulted was an entirely new instrument, one that did not produce sounds from ∫-shaped holes in a hollow wooden body. An electric guitar is, in a sense, a powerful input device to control the electric amplifier and speaker which emit the

159

sounds. Les Paul was a pioneer among the many innovative electric guitar players who produced music the likes of which human ears and brains had never encountered before.

The electric guitar was the main weapon for a musical revolution in the 1950s and early '60s led by Chuck Berry, Bo Diddley, Carl Perkins, Duane Eddy and Dick Dale, and country pickers like Chet Atkins, who played on many of Elvis Presley's early recordings. Bo often played a guitar that hardly even looked like one, with its rectangular shape that shouted, this ain't no acoustic guitar. He didn't play it like one either. Rhythm was the heart of his sound.

The musical explosion of the 1960s was driven by musicians exploring the new sounds they could wring from their instruments, "guitar gods" like Jimi Hendrix, British blues-rockers Eric Clapton, Jeff Beck and Jimmy Page, and electric blues masters Muddy Waters, John Lee Hooker and B.B. King. Bob Dylan shockingly went electric. For The Beatles and producer George Martin, and many other artists, the recording studio, with its multiple tracks and electronic devices to control and alter sound, became an instrument in its own right. "Sergeant Pepper" taught the band to play it.

One of the most exciting and experimental bands of that era was The Velvet Underground. They brought elements of Fifties rock 'n' roll, contemporary avant garde and futuristic electronics into a sound that rocked relentlessly, with uniquely poetic and often dark lyrics. The group, initially promoted by Andy Warhol, was formed by Lou Reed, an aspiring if offbeat songwriter who grew up on Long Island, NY, and John Cale, a Welshman with a classical music background who came to the States with a fellowship funded by Leonard Bernstein.

The Velvet Underground
(clockwise from bottom left:)
Nico, Lou Reed, Sterling Morrison, John Cale, Moe Tucker

Lou was lead vocalist and basically played rhythm guitar, but took off on wild flights of soloing and feedback. John pulled unearthly shrieks from his electric viola, when he wasn't beating an electric piano into submission. Sterling Morrison entwined his rockabilly and country-inflected guitar playing with Lou's burning guitar runs. Maureen "Moe" Tucker, the former key punch operator, laid down a simple throbbing beat evoking the primitive heart of rock 'n' roll's rhythmic roots. In the band's initial incarnation, the chanteuse Nico contributed her mournful and ironic alto voice to several numbers. It was the apotheosis of electric music.

The first time I saw them play I didn't know it. I was at an event presented by the literary magazine *The Paris Review* in April 1966 at the famed club The Village Gate. The place was thronged with New York's literati and glitterati. Down in the lower level of the club several rock bands played. I didn't hear them get introduced, but the last band to go on was hypnotic, strobe lights turning the churning bodies on the dance floor into flashing erotic tableaus. Out of the corner of my eye I spotted a master of acoustic music coming down the stairs, Frank Sinatra. With a look of shock on his face, he suddenly paused midway, and then retreated back upstairs. Not his scene, man. But he had seen the future, and so had I. The sound of that band reverberated in my mind for months.

I only learned who they were a year later, when I saw them on the cover of their first album, designed by Andy Warhol with a peelable banana on the front. I was living in Cambridge at the time, and a few weeks later, after nearly wearing the grooves off their record from repeated playing, saw them play at a club called The Boston Tea Party, a psychedelic temple of rock. Built as a Unitarian meeting hall in 1873, its boxy shape and high ceiling was well-suited to rock acoustics. When the Velvets did a twenty-minute rendition of their classic "Sister Ray," the whole room vibrated, the people and structure. You had to be there to experience the electricity they generated.

By a strange twist of fate, I became the manager of the Tea Party and produced rock concerts in western Mass. as well. I got to know the band personally and promoted many of their shows. The VU took electric music into new dimensions. I was glad I could help them along that journey, a hard road with little financial reward. John Cale left the band in 1968, replaced by Doug Yule. But when Lou Reed left in 1970, it was over. Many years later, I saw Lou backstage after a show in western Mass. He introduced me to his new bandmates and told them of my involvement with The Velvet Underground, saying about me, "We could not have survived without him."

The band existed for less than five years and never had a hit record. Yet they left indelible footprints on the evolution of electric music, and inspired musicians for decades to come. Rolling Stone magazine called them "the most influential American rock band of all time."

Poster promoting shows by The Velvet Underground at New York nightclub Max's Kansas City, 1970

Lou Reed's guitars and amps in "concert" at Mass MoCA

Laurie Anderson and her electric violin

In a 1978 interview with the Los Angeles weekly newspaper *Open City* ("All Tomorrow's Parties", issue #78) Lou explained what he'd experienced with the VU playing a guitar with a fuzz pedal: "You go like this on the fuzz, daaaaaaaaaa – and what happens is that you don't get just one note like a guitar, you may get eight notes, like daaaaahhhhheeeee.... You start hearing some really strange things.... We used to call it the Cloud, and like, on certain songs, we used to consciously enter the Cloud and you just hear all these funny things. They're not you, but you know they're being caused by the guitar, right?"

In 2019 I attended a "concert" caused by Lou's guitars six years after he died, "Laurie Anderson Presents: Lou Reed Drones," at the Massachusetts Museum of Contemporary Art. Entering a large performance space, I saw seven of his guitars leaning against seven of his amps in a semi-circle, the room resonating with the loud drone of feedback.

The event was produced by his former guitar tech Stewart Hurwood, who "sets a soundscape foundation, manipulates and interacts with the guitars; the feedback is altered by the location and movement of people present in the space." Hurwood explained that "what we're dealing with is interference patterns so that guitars are going to be playing against each other." (*RNZ Music*, 3 March 2020)

It was an electrifying night. Lou's widow Laurie joined in, playing an electric violin unlike anything Stradivari crafted. Electric music continues to evolve. Synthesizers swing. Rap is rhyming rhythm. From the Stone Age to the rock age, the beat goes on. As Plato wrote in *The Republic*, "Rhythm and harmony find their way into the inward places of the soul."

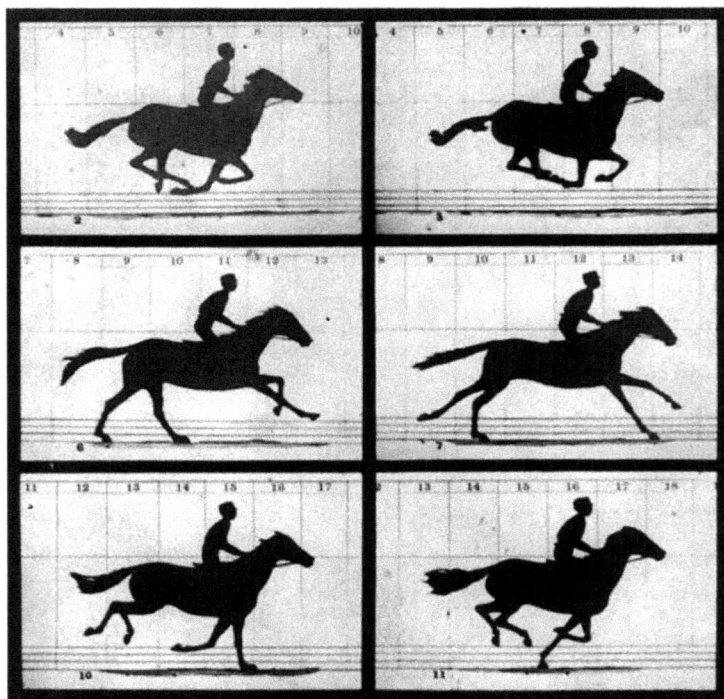

Eadweard Muybridge, "The Horse in Motion"
(partial sequence)

THE MIND'S EYE

Just as electricity let humans hear sounds never heard before, so too did it let us see things never seen before. It had long been debated whether all four of a galloping horse's hooves were ever airborne simultaneously, or whether at least one hoof was always in touch with the ground. The problem was that human eyes and brains could not process a horse's fast-moving legs quickly enough to tell. In 1878, the business tycoon, racehorse owner and former Governor of California Leland Stanford got the photographer Eadweard Muybridge to document a horse's gait and finally settle the question.

Muybridge had been capturing animal locomotion with a series of stop-action photos. He set up twelve cameras and a backdrop at a track on Stanford's horse farm (the site of the university he founded seven years later). A horse running past them tripped wires connected to an electromagnetic circuit which triggered the shutters. There it was in the photo sequence revealed at last: the horse in full flight.

Muybridge went on to produce thousands of images of animals and humans in motion, what he described as "an electro-photographic investigation" of movement. But not satisfied with being limited to photos, in 1880 he invented the zoopraxiscope, which projected sequences of still images to crudely animate them. In 1889 Thomas Edison invented the Kinetoscope, a box the viewer peeked into to see short films. In 1896 he debuted his version of a true theater motion picture projector, the Vitascope. These devices used a moving series of still picture frames, which occurred rapidly enough that the human brain interpreted what the eye saw as continuous movement, the basis of movies and video today.

But Edison was, uncharacteristically, not first when it came to projecting films. The French brother team of Auguste and Louis Lumiere had produced screenings a year earlier. Newsreels and actualities became popular forms of early films. Their potential as a profitable entertainment medium was demonstrated in a film of a heavyweight championship boxing match in 1897. Running 100 minutes, it was the longest film made to date. Its box office ticket sales exceeded the live gate at the bout. Show it and they will come.

Motion pictures evolved to be in color, on larger screens, in higher resolution, with high-fidelity stereo sound, and displayed digitally rather than by projectors. In April 1968 I saw a landmark of cinematic presentation, the uncut version in Cinerama of Stanley Kubrick's *2001: A Space Odyssey*, a film about human evolution. His depiction of violent cavemen has been disputed, but the film is an allegory, not a documentary.

Television is another visual medium based on depicting motion with still frames, which are scanned onto the screen. Viewing video on a large 4K HD monitor with surround sound in the comfort of home rivals if not surpasses going to movies in theaters. And coming into focus are new display technologies like AR (augmented reality) and VR (virtual reality), with holographs at home on the digital horizon.

The point of all this is not simply about eye candy. Our brains are being trained to see these images and to savor them. When a very early film depicted a train heading toward the audience, people panicked and fled in fear. In 1953 I took my first date Gloria to see the first 3D feature film, *Bwana Devil*.

When a lion jumped toward you or a spear seemed to hurtle through the screen, you reflexively ducked, and some people even screamed. But seeing *Avatar* or *Ice Age* many years later in 3D, you looked on in wonder and even laughed in enjoyment at the spectacular effects. We got the picture.

To see how our perception has been trained, compare an old TV show or movie with one today. It's noticeable how much more slowly images cut from one to another back then. This is especially pronounced with TV spots. Search for "1950s TV commercials" on YouTube and notice the leisurely pace of the editing. TV spots today hit you with one image after another at a frantic pace. Bam bam bam.

We've become practiced at discerning fast-cut images, on screen for a fraction of a second before another appears. Our brains have learned to process them. If you could go back to the 1950s and air a TV spot produced in today's rapid-fire style on "The Milton Berle Show," viewers would be confused, not sure just what they're seeing. The quick cuts would make them uncomfortable, not a very good way to sell the sponsor's Buicks. But as we now see, it works today.

Electricity has not only enabled us to capture and display moving images, but to create ever-more complex and realistic computer-generated animations and effects. We can view the smallest microscopic objects and peer deep into space at astral bodies so far away that what we're seeing, when their light finally reaches us, occurred thousands or more years ago. In an upgrade to Muybridge, M.I.T. professor Harold Edgerton ("Papa Flash") pioneered using strobe lights to photograph athletes in motion, hummingbirds flying, and bullets in flight. Our electrically-aided eye can catch a speeding bullet.

Nude by Nadar, 1860-61

SEX TECHS

Electricity has changed the ways people can experience sex, perhaps accompanied by low lighting and music. It has enabled the production, presentation and consumption of erotica and pornography, and the manufacture, distribution and use of orgasm-inducing devices, a.k.a. sex toys.

Since the first photograph taken about 200 years ago, the camera has been used to portray nudity, especially the female body. The early French photographer Nadar captured many leading public figures of his day, but also turned his eye toward the nude female figure. Muybridge simulated animation with his sequences of still photos of nude men and women in motion. The intention of these artists and inventors was not titillation, but they brought the depiction of nudity to a far wider audience than the ladies and gentlemen strolling among the paintings and sculptures of nudes in art museums.

The introduction of true motion and film projection was a game changer. In 1896 an 18-second film called *The Kiss*, with two actors smooching rather chastely, was a sensation, drawing large audiences while fueling demands for film censorship. With better quality and the addition of sound, movies got even hotter, while censors grew hotter under their collars. The Catholic Legion of Decency, formed in 1934, inaugurated film ratings, from "A" okay" to "C" no way.

The first film condemned with a "C" was *Ecstasy*, starring Hedy Lamarr. It showed her swimming in the nude and then chasing the horse which ran away with the clothes she had draped over it. Her face was shown in close-up simulating an orgasm. But when Hedy had her clothes on, she was the brilliant inventor of a radio guidance system for torpedoes based on spread spectrum and frequency hopping technology, still used in telecommunications today.

Hollywood reluctantly self-imposed a Production Code to control sex on the screen, motivated by the desire to keep the government's hands off. But producers and directors found ways to slip steamy stuff in, knowing that audiences wanted to see it. The Code lasted for decades until replaced with the now-familiar G, PG, PG-13, R and NC-17 rating system. It had the unintended effect of alerting viewers as to where to find the more erotic fare.

Sexual content in European films so outstripped scenes in Hollywood productions that they were called "art films" and screened in independent "art" houses. The manager of one such theater near Cleveland was convicted in 1960 of possessing and exhibiting an obscene film in violation of Ohio law: *The Lovers*, directed by Louis Malle and starring Jeanne Moreau as a woman having an adulterous affair. The case went to the U.S. Supreme Court, where the film was ruled not pornographic and the conviction was reversed in a 7-2 vote.

The Court was hopelessly conflicted as to the rationale for its decision, issuing four separate opinions, including a dissent by the usually liberal Chief Justice Earl Warren. The opinion famously remembered from the case was that of Justice Potter Stewart. In ruling the film was protected by the Constitution because it was not hard-core pornography, he demurred from defining pornography but wrote: "I know it when I see it."

Plenty of people were seeing it, in movies from an under-ground porn industry that surfaced its productions in seedy theaters in Times Square and elsewhere. Their business boomed with the advent of home video. No longer burdened by the cost of film stock and the equipment needed for film production, porn producers set up makeshift studios in suburban homes in places like the San Fernando Valley in southern California. They cranked out cheaply-produced videos that were sold in "adult" stores in strip malls and downtowns across America, a clandestine but flourishing business. The censors were usually nowhere in sight, perhaps gone extinct like the Neanderthals, but under-the-table money may have kept the law away from some stores, a necessary but affordable cost of doing business.

But the biggest boom was yet to come when technology eliminated hard media and physical distribution. Cable TV had been airing boxing matches on a pay-per-view basis since 1960, but in the 1980s rolled out a multichannel PPV movie service. Among the offerings was soft core porn, much less explicit than videotapes but much more convenient. Playboy TV was launched, and to keep up with its competitors got harder and harder. Cable executives didn't like talking about it, since they were licensed to operate in many conservative cities and towns, but sex was a big source of PPV revenue.

Playboy and the flood of "skin books" which followed in its wake were the products of electric printing presses. But it was the internet which blew the doors off porn. It was everything goes all the time, accessed with a fast internet connection. A new phenomenon emerged: porn produced by women for women, for men, for everyone. Some former female porn stars, past their nubile prime, saw new opportunity behind the camera, drawing on what they learned about the business from acting in front of it.

This was part of a trend of women taking control of their sexuality. Porn had long been primarily a medium for men. The women who appeared on camera made it possible of course, and they did it for money or other motives. Much is made of the male gaze, but its counterpart, the female display, became more explicitly obvious, and more explicit. Women had long adorned and exposed their bodies, part of the sexual selection process Darwin observed, although it's hard to infer from old skulls and bones to what extent this was true of ancient humans. We know of no erotic cave paintings, unless some archaeologist has kept them under wraps. But with smartphones, women can self-produce, star in and sell their own porn. Outlets to monetize their videos online have made this a lucrative side hustle, for some a full-time occupation.

The electric vibrator was invented in 1878, but it is disputed whether it was intended for pain relief or, secretly, for sexual release. By the sexual revolution of the 1960s, it had come out of the closet, or perhaps I should say out of the drawer, as a device used by women to achieve orgasms. The pandemic stimulated explosive growth in the sales of electric-powered sex toys to women home alone. They have become mainstream appliances, tested and reviewed by writers in the "Wirecutter" consumer products feature of The New York Times, and sold by websites like Amazon and upscale department stores like Saks Fifth Avenue. May I help you, madam?

To Darwin, sex was solely for propagating the species by transmitting traits to progeny. For humans today it can also be casual, recreational and self-satisfying. Remote by text, voice and images sent by smartphones. Or even with an AI as your partner. Electricity gave sex a new charge.

ALTRUISM

One of the more puzzling aspects of human behavior, from an evolutionary perspective, is altruism. How does doing good fit into a model in which creatures struggle to survive, often at the expense of others? I wrote earlier about the parallels between Adam Smith's view of the economy and Darwin's view of evolution. Smith believed that if each economic entity pursued its self-interest, that would lead to the greatest good, but it's not obvious just how it would.

If all commercial fisherman work to maximize their catch, by whatever means necessary, a la Smith, the result could well be the depletion of fish stocks. That is exactly what happened in the late 20th century in Gloucester, Massachusetts, where I lived for many years. It is the oldest fishing port in America, adjacent to some of its richest fishing grounds. The situation was exacerbated by the giant Russian floating fishing factories which trawled in nearby international waters, in effect vacuuming fish from the seas where Gloucestermen also trawled. That was in the Russians' self-interest, to maximize their catch and leave.

This conundrum was first conceptualized by the English writer William Forster Loyd in 1833 (between the time Smith published his theory in 1776 and Darwin his in 1859), and later termed "the tragedy of the commons" by Garret Hardin, writing in Science Magazine in 1968. The tragedy is that people with access to a public resource, "the commons," who use it for their unfettered self-interest, may well deplete it.

Altruism restrains people from their more selfish instincts. Efforts to explain it in Darwinian terms have not been entirely satisfactory. The answer I believe lies in that aspect of human development which Darwin did not take into account, and which has only been widely recognized by scientists in recent decades: cultural evolution. Cultural and social taboos evolve to hold most people back from their worst predatory instincts, for the benefit of the commonweal. We can feel good by doing good to safeguard the commons that is planet Earth.

In the case of the fishermen, they altruistically albeit reluctantly agreed to regulation of the fishing industry which imposed strict limits on catches in national waters to allow the fish stocks a long period of time to replenish. Many fishermen sold or abandoned their boats and left the trade, a necessary result of rejuvenating the commons. It was up to Washington and international bodies to deal with the Russians.

The concept of the commons is not limited to commerce. In the Declaration of Independence, the founders committed their "lives, fortunes and sacred honor" to the establishment of a democratic form of government, our national commons. Abraham Lincoln, Franklin Delano Roosevelt and countless others, famous and not, have worked to preserve and improve it. Many died to protect it, soldiers and civil rights workers.

That is the commons we all share as Americans, a democracy based on the values of "life, liberty and the pursuit of happiness." At its foundation is the peaceful transfer of power, the altruistic principle that losers of elections willingly step down in the interest of preserving the commons. One man decided recently that his self-interest must come first by denying the legitimacy of the 2020 presidential election. His actions then and now do not make America great. They lessen it by undermining the commons of our democracy.

E.I. -- A.I. -- OH!

Just as fire enabled bigger brains, electricity enables a new product of evolution, extended intelligence (EI): the ability to store, access and process information outside the brain, at speeds far greater than the unaided brain is capable of.

Early humans scratched lines onto an object or surface to record an amount of something. Cave paintings were not only an expression of their aesthetics, but portrayed a particular hunt or a hoped-for one, serving as a lesson plan for future hunters. They placed dots, lines and other simple symbols on paintings of animals which have been found to correlate with their mating and birth cycles according to the lunar calendar.

In China and Europe during the late Stone Age, symbols were carved onto turtle shells and wooden tablets. In Mesopotamia by about the fourth millennium BCE, clay tablets were marked by styluses as a method of record keeping, and the cuneiform language which emerged was used by the Sumerians and others in the ancient Middle East. Similarly, the Egyptians developed hieroglyphics as the written and highly symbolic form of their language. New means of recording and transmitting information – scrolls, books, etc. – emerged over the centuries before the Electric Age, but these were all produced by mechanical means and their content was fixed. You had to hold an object or stand in front of it to extract its meaning, an often-laborious process.

With EI, information can be transmitted, received and interacted with in real time over vast distances. Electronic memory in computers and smartphones has changed how we deal with all kinds of information. As a simple example, we used to have to memorize phone numbers, or write them down somewhere. Then we rotated a little disk with holes corresponding to letters and numbers, or pushed the appropriate sequence of buttons on a phone, to "dial" the phone number. Now our phones remember those numbers for us, and by touching or saying a name to select the person we want to call, the phone "knows" how to do the rest. The intelligence to store and utilize the phone number data resides in the device, not in our brains. I used to know dozens of phone numbers, besides those I looked up in my Rolodex or business card file. Now I know only a half dozen or so, including 411 and 911. It's not that I've gotten dumber. Phones have gotten smarter, our extended intelligence more powerful.

People used to think you were smart if you had a lot of facts and figures at your fingertips, colloquially speaking. Now they actually are at your fingertips, with the tap of a keyboard or screen, or the click of a mouse. It's not what you know, it's what you know about how to find out what you want to know. Those smarty-pants with all the answers can go compete on Jeopardy. With my computer I know more, or more accurately am able to know more, than they ever will.

Earlier I wrote about how we live bathed in EMR, the electromagnetic spectrum. We also live in what has been called the "infosphere." This is a term coined in 1970 by the economist and philosopher Kenneth E. Boulding, and refers to the informational environment surrounding us. As the philosopher Luciano Floridi interpreted it in 1999, it encompasses "all informational entities (including informational agents), their properties, interactions, processes, and mutual relations." Cyberspace is but a part of it.

But while the infosphere is omnipresent, each of us accesses it and uses it in different ways and to different degrees. We each experience it uniquely, as an extension of our individual brains. I call that extension the "exebrum," the virtual part of our brains which exists external to our central nervous systems and is the source of EI. It is a distinctive characteristic of *Homo electric*. It won't show up in scans or endocasts of our skulls. But it is there, connected to our brains through our internal wiring.

I wrote earlier about Professor Joseph Henrich's view that "our ability to learn from others" and amass "larger bodies of adaptive knowledge" has given rise to what he calls "our collective brains." Through the exebrum we can access and utilize this common knowledge of our collective brains. Henrich further argues that the collective brain has been more responsible for innovation than the lone inventors of legend. While I agree in the larger sense of how innovation arises and spreads through society, I think that a qualification is in order.

Brilliant individual innovators are still to be found, like Steve Jobs, central nodes on a collective network, but they work closely with collaborative teams of brains. Macintosh was Steve's vision, and he had a talented team of engineers, designers and programmers to implement it. Look inside the plastic box housing the electronics of the original 128k Mac, and you'll see their names inscribed, like handprints left on an ancient cave wall saying "we were here." With computers and the internet, collaboration now occurs across the barriers of space and time in a way it could not, say, with the development of the atomic bomb at Los Alamos under the leadership of the innovator J. Robert Oppenheimer.

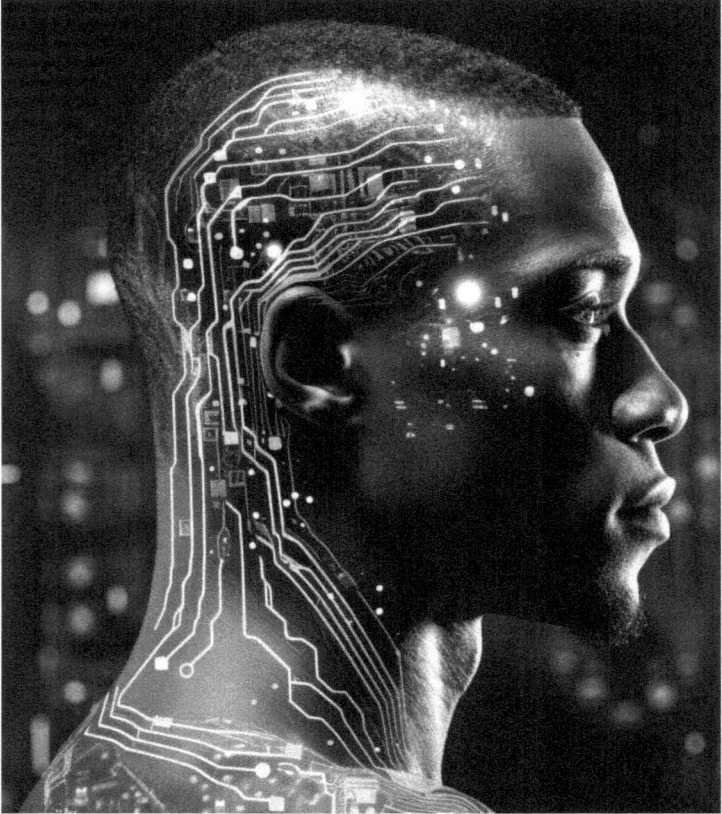

A conceptualization of the "Singularity" generated by AI

Rouzbeh Yassini-Fard, introduced earlier as the father of the cable modem, is one of several innovators who helped make the internet into the critical connector of our brains. He stresses the importance of "ubiquitous, low cost, end-to-end broadband connectivity" as a matter of social equity and economic opportunity. He notes that half the people of our planet are unserved or underserved in accessing the internet. I would add that because the internet is essential for a fully developed exebrum, half the people of the planet lack the extended intelligence that enables them to evolve as *Homo electrics*. We need to connect them to allow them to fulfill their human potential.

Some futurists like Ray Kurzweil predict (and even welcome) what they call the "Singularity," the future point in time when technology evolves irreversibly beyond our control and we are subsumed by it. But when humans learned to harness electricity, it quickly became uncontrollable because it was in too many hands. It was certainly irreversible, because as I noted earlier, no one wants to give it up.

The fear of technology taking over was memorably represented allegorically in the TV series "Star Trek: The Next Generation." The Borg were cybernetic aliens interconnected and controlled by a hive mind. They conquered other species by assimilating their technology and turning them into drones augmented by cybernetic components, making them part machine. They ominously told their victims that "resistance is futile."

I bring up these fictional creatures because of the real-world fear of artificial intelligence, a very powerful form of EI. AI bots are able to use the massive amount of information they have been trained with (instructed to absorb), or found on their own online, to perform tasks that only humans once could, like writing research papers and generating complex graphics. They do so at the bidding of humans, but then operate on their own to produce the requested result.

Let's not forget that computers were invented in the first place to perform tasks that were beyond the ability of unaided human minds, like deciphering German military codes during World War II, or modeling thermonuclear explosions. But AI is far more powerful. Computer hardware and software have evolved to such a degree that some now fear AI may be able to act in ways contrary to human interests, like the Borg, even threatening our continued existence by developing a "mind" of its own, known as artificial general intelligence. It is hotly debated whether AGI can exist, not as an extension of human intelligence like AI bots, but totally independent of us, and whether AGI truly can be equal to human intelligence, or even superior to it.

Such concerns led hundreds of computer scientists and others to sign a petition calling for a six-month moratorium on AI development, a plea to desist while the federal government promulgated laws and regulations to oversee it and mitigate its dangers. Six months, a year, five years, how long would it take for the government to adopt such a regulatory regime, without stifling AI altogether? It has never effectively regulated newly-emerging technologies, like the internet, other than those under its direct control, like nuclear power.

In overseeing AI, could Congress produce anything useful? Federal and state governments will perhaps do best by doing the least, certainly not doing anything hasty under an artificial deadline and political pressure. In the meantime, of course, people who didn't sign that petition were free to go about their AI business, motivated by a financial bonanza that may dwarf any before it. Other than raising awareness of the problem, and fanning fear of AI, the petition went nowhere. Desistance is futile.

The debate about AI will continue, flaring up with each new breakthrough or new threat in AI capability. We will have to work together to reap its benefits while minimizing its dangers. It is inevitable, our evolutionary destiny, that we will move ahead to develop AI, improving this tool as humans improved all their tools since inventing the stone hand axe.

Ironically, we will need AI to mitigate another existential threat to our species: climate change. To maintain the sustainability of the ecosystem of Planet Earth, eventually we will have to manage it as we would manage any complex system, but on a vastly greater scale than anything we've ever done before. We will need AI to do it.

It may be necessary, in managing the planet's ecosystem, to impose limits on energy use and emissions, temporary or long-term (AI is itself a massive energy hog). But technology alone will not make it possible to protect our planet. It will require unprecedented cooperation by the leaders of all nations, and their citizens. Cooperation enabled *sapiens* to build civilizations together, despite the aggressive territoriality of the species (manifest today, for example, in the obsession over the U.S. border). It will be the ultimate test of *electrics* to achieve a far higher level of cooperation than humans ever have before. That is our evolutionary future, our road to survival which we must travel together.

PART SEVEN

A Planet Electric

Is it a fact –- or have I dreamt it – that by means of electricity, the world of matter has become a great nerve, vibrating thousands of miles in a breathless point of time?

-- *Nathaniel Hawthorne, author*

Entrance to the fire lookout tower on Massaemett Mountain

A

In the summer of 1970, my future wife Jan and I lived for a month in a simple one-room cabin on Massaemett Mountain in western Massachusetts. It had electricity, but no phone, radio or TV. We spent part of our days writing fragments of poetry about love and sex, the past and the future, time and space. A footpath from the cabin led uphill to an old stone fire lookout tower with a panoramic view. We climbed its stairs every evening to watch the sun set beyond the rolling hills.

The cabin's "library" had two books, one of no interest. The other, published in 1930, was the first-ever autobiography by a Native American leader, the Crow chief Plenty Coups, named for his many acts of bravery. He told of a vision quest in Montana's Crazy Mountains, where he foresaw the disappearance of the buffalo, the staple of life for Plains Indians.

Alone with our thoughts in the cabin and atop the tower, we envisioned the future disappearance of the "buffalo" of our modern way of life: oil. And we imagined the end of nation-states and the emergence of a globally interconnected human consciousness. With my background in math, I used the letter "A" to represent this unknown, this global consciousness. It was visually symbolic as well, with the two opposing sides rising to unite, held together in the middle by a common bond.

We wrote:

> We are all creating A.
> A is creating all of us.
> America. Russia. China.
> They all end in A.

We had no idea how such a crazy vision could come to pass.

After we left the cabin, Jan and I assembled our typed tidbits into a self-published little volume called *Sunsets Are Red White and Blue*. The title evoked beautiful evenings on the tower watching the sun go down behind an American landscape. But it also spoke to the sunset of America as a nation, at some indeterminate time in the future.

It was not a time I would welcome. Despite being part of the 1960s counterculture, I had always considered myself a patriot. As a student at Harvard Law School, I studied and admired the U.S. Constitution, created "in order to form a more perfect Union," imperfect though that union was and is.

But a foreshadowing of that time did appear, on January 6, 2021. The President of the United States used social media to call together a mob, which he then incited to storm the Capitol Building in an attempt to block certification by Congress of the electoral votes which decide a presidential election. Their insurrection was driven by a "conspiracy theory," a paranoid delusion that the election had been stolen, despite all evidence to the contrary. They rioted to prevent the peaceful transfer of power, the bedrock of our democracy. Call that president "T" and call his followers "TAnons." Understand him as a "sapiopath" in whom the very worst aggressive tendencies of *Homo sapiens* manifested themselves, and still do.

Most Republicans in Congress refused to certify the election and still claim (at least publicly) that it was rigged. Many of their GOP counterparts at the state level are working to undermine the electoral process. Many voters identifying as Republicans cling to their fantasy about a stolen election. With delusion so rampant, how can the United States continue to function as a viable society?

It is not the purpose of this book to suggest political solutions to our current situation. But the issue of regulating social media, characteristic products of the Electric Age, does merit a few words here. They have been powerful forces for both human interconnection and deep division.

In the aftermath of January 6, both Facebook and Twitter banned "T" from their services, while Amazon, Apple and Google kicked the right-wing social media platform Parler off their servers. That was quite a lineup of internet heavyweights acting loosely in concert, although "T" and Parler found new homes online. The question is, however, who or what should be able to take measures like banning a public figure like "T", labeling a social media post as misleading, or removing deep fake porn? Is it solely up to the social media companies to decide? Or is there a public interest that they must answer to?

The power of these huge companies is attracting increasing scrutiny. But we can't just say big is bad, like in the days of Teddy Roosevelt and the trustbusters. To a degree that we've never seen before, bigness (the new term is "scale") is inherent in such internet companies. There have been calls to break them up, but that is not an easy solution. Who wants to be on a social media service whose reach is curtailed?

Social media are unique to the Electric Age, and do not fit comfortably under the freedom of speech and press provisions of a First Amendment adopted in the 18th century. Their owners are not, as they claim to be, like newspaper publishers with the unfettered right to decide what information is "fit to print." They should be subject to some degree of public accountability, yet without inhibiting their ability to operate and innovate. It's a fine line to walk, but walk it we must.

EARTH RISE

When astronauts first circled the moon and looked back at the Earth, it was a revelation. To see our planet rise behind the moon was to feel a new awareness of Earth as the home to all human beings, to all known life. A planetary awareness.

In 2015, at a United Nations conference, 196 nations adopted the Paris Agreement on Climate Change. Its aim was to cap the increase in the planet's temperature at 2° C above pre-industrial levels, and preferably at 1.5° C. This action was an unprecedented historic event, the world coming together to address the existential threat of global warming.

The Obama Administration played a key role in making it happen. But when Donald Trump became president, he took the steps allowed under the Agreement to withdraw the United States. On the first day of Joe Biden's presidency, he issued an Executive Order for the U.S. to rejoin. On the first day of Trump's second term, he pulled us out again, and has cut U.S. climate change and renewable energy programs.

This exposes the fatal flaw of the Agreement: participation is voluntary, subject to the whims of *sapiens* potentates like Trump, Putin and Xi. Even if a member country votes for measures put forth by the body of members, it is only voluntary that it implement them. The Agreement has no mechanisms to enforce its own goals; countries were not willing to surrender any of their sovereignty to a global body when they negotiated the Agreement. Global temperatures now regularly exceed the 1.5° C goal for extended periods. I believe regrettably that the Paris Agreement, in its current form, will fail to control climate change.

Eventually, evolution will provide a solution, as *electric* leaders emerge, guided not by sovereignty but by their planetary awareness, by their willingness to invest a global body with the authority to enforce measures necessary to protect our planet. As the average temperature continues to rise, as weather disasters become more frequent and more fierce, we will be forced to evolve, to come together as one world in order to survive. Let's hope we do so in time.

Since hunter-gatherers formed the first agricultural settlements 12,000 years ago, we have lived in ever larger societies. This required a level of cooperation unique to humans, but also large armed police and military forces to maintain order. Electricity made feasible today's towering cities of millions.

Germany played a central role in two World Wars as a powerful militarized country, yet it was not until 1871 when the princes ruling dozens of independent Germanic fiefdoms united to form a single nation-state. That led to dire consequences, needless to say, but today Germany is a member of the European Union, a major force for world stability and peace, and a leader in reducing the use of fossil fuels and adopting renewable energy.

With Britain's exit from the European Union, and the rise of authoritarianism in many nations, it may seem that we have come to the end of the process of globalization. We have not. These are just bumps on the long road of evolution. Why assume that in the future, the world order will not evolve from its form today? It has before. It will again.

The real obstacle to global cooperation, and ultimately unification, is *Homo sapiens*. As a species they have been motivated by a strong sense of territoriality. Male *sapiens* are highly protective and controlling over what they perceive to be their territory and their property, which in their minds often includes females. The emergence of women asserting themselves in the political, economic and creative spheres, and in the control of their bodies for reproduction and sexual pleasure, is not merely a sociopolitical trend. Feminism is not just an ideology or a movement. It is a manifestation of evolution.

We are evolving from only interacting face-to-face to doing so remotely, from unconnected to interconnected, from mechanical to electrical, from patriarchal to equiarchal, from resource exploitive to conservationist, from territoriality to planetary awareness, from *Homo sapiens* to *Homo electric*.

Note that word "equiarchal." It is said to have been coined by the Norwegian sociologist Dr. Johan Galtung, founder of the academic discipline of peace and conflict studies. He intended the word to be a cross between hierarchy and anarchy, referring to an equal relationship between nations. But I am using it in a different sense, to refer to an equal relationship between the sexes, neither patriarchy nor matriarchy. Equiarchy. A characteristic of *electric* culture

Even those *sapiens* who are adept at using the technologies of the Electric Age, like social media, will not fully evolve into *electrics* until their consciousness evolves. Being an *electric* is to feel our oneness with all human beings, with all living things, and to embrace our stewardship of Planet Earth. As *sapiens* fade and *electrics* arise, we will come together.

I'm hardly the only one to imagine a world united as one. John Lennon famously did. Pierre Teilhard de Chardin (1881-1955) was a paleontologist and French Jesuit priest who believed that individual human thought would eventually merge "to form a global consciousness and self-regulating super-organism called the Omega Point" (David Sloan Wilson, *This View of Life*). Wilson wrote that the word "organism" can refer to large entities "such as a human society or a biological ecosystem." Or to an entity such as A?

The idea of the earth as a self-regulating complex system was proposed in the 1970s by chemist James Lovelock and microbiologist Lynn Margulis. Their "Gaia hypothesis" was named for a goddess in Greek mythology who personified Earth. In Marshall McLuhan's "global village," media provides the interconnectedness, a concept he formulated decades before widespread adoption of the internet. He said that "the electric age...established a global network that has much the character of our central nervous system." Internet pioneer Rouzbeh Yassini-Fard says: "Eventually, we'll learn to use the network to improve human life and extend the planet's survival. That's the ultimate goal."

Responses by the scientific establishment to "one world" theories have often been critical, with some justification when you dig into their details. But in a larger sense these ideas, although imperfect, express a growing belief in our connectedness as humans and the unity of our planet. They seek to explain how such a superorganism might come to exist, as inevitably it will with the emergence of *Homo electric*. It is a unity not merely of political and economic entities, but of an interconnected consciousness.

The concept of consciousness has long puzzled scientists, philosophers, theologians, and just about anyone who thought about it. What is it? How does it work? Where does it come from? No one has been able to say for sure. And yet we are conscious of our consciousness. To paraphrase Justice Potter Stewart, "I know it when I sense it."

The French philosopher Rene Descartes put it another way, when he famously said, "Cogito, ergo sum." I think, therefore I am. To think "I am" is to be aware that "I am." We know the brain can think such philosophical thoughts, can enable an author to write a book, can conclude that it's a nice day because it's aware the sun is shining. We think it is somehow involved in generating and sustaining consciousness, but we do not know exactly how. We do know that the brain and nervous system run on electricity. When a person's brain shuts down with death and its electrical activity ceases, so does their consciousness. I'll leave it to others' beliefs as to whether consciousness or a "spirit" continues in another form.

We can say that consciousness is awareness of one's self, of the external world and our place in it. As we evolve, that awareness, that consciousness, is also evolving. Electricity connects us. Through the internet we share more than information, we share our feelings, our ideas, our dreams. It can enable us to achieve a higher level of cooperation. Along with the devices connected to it, the internet is becoming the external electrical counterpart to our internal electrical network. It is giving rise to an interconnected global consciousness, an *electric* consciousness woven, as Ada Lovelace might have said, into the fabric of what it means to be human.

We are the weavers of our lives, of our ongoing evolution in the Electric Age, of the future of Planet Earth. Natural selection may still play a part, but it is the actions we take which will ultimately guide the course of the evolution of the human species, beyond *Homo sapiens* to *Homo electric*.

WORDS FROM THE WISE

To Charles Darwin, evolution was a passive process, the invisible hand of natural selection determining who was fittest to survive, whose traits would be inherited by future generations. So it was as well to Gregor Mendel, Ronald Fisher, James Watson, Francis Crick and Richard Dawkins.

But many evolutionists now recognize the impact on the human genome of the sigmasphere, and especially of cultural evolution, in which we play an active role and interfere with the operation of natural selection. Recognition of climate change, as a peril to our planet and all living things, is fundamental to *electric* culture. Were the polar bears to go extinct, as they may well, it would not affect us in our everyday material lives. But it would deeply affect us psychologically.

David S. Moore cites recent research in epigenetics which "has started to show how our molecular biology influences our psychological states *and* how our psychological states influence our molecular biology" (*The Developing Genome*).

Joseph Henrich writes that "the central force driving human genetic evolution… has been cultural evolution," which gave rise to "our collective brains." He says that "the degree of social interconnectedness is very powerful in generating cumulative cultural evolution" (*The Secret of Our Success*).

David Sloan Wilson says the technologies which connect us "are capable of furnishing the earth with a planetary brain." He adds "we must be the navigators, consciously evolving our collective future" and "mindfully direct the process of cultural evolution toward planetary sustainability" (*This View of Life*).

We've got the whole world in our hands.

A NEW SPECIES OF HUMAN

"There is a grandeur in this view of life," wrote Darwin about his theory of evolution, "endless forms most beautiful and most wonderful have been, and are being, evolved."

No form is more wondrous than that most unique animal, *Homo sapiens*. It emerged 200,000 years ago or more, and flourished as the sole survivor of a complex hominin lineage. Then, less than 200 years ago, humans learned to use electricity, and a new species began to emerge, differing from their pre-electric predecessors in significant ways:

- they use electricity rather than fire as their primary source of energy to support their existence and development; the electrified world is their natural habitat;
- they can store, access and process information by using extended intelligence, a virtual external part of the brain, and its rapidly advancing form, artificial intelligence;
- they are interconnected and able to interact remotely in in real time, express their emotions electronically, and communicate to millions simultaneously;
- they can see things, hear things, do things, know things and imagine things never before possible;
- they can experience the past via recorded media;
- they share an exponentially greater body of knowledge, experience, tools and problem-solving techniques and technologies (collective brains);
- they can leave the surface of the earth to travel through the air and deep beneath the seas, and live and explore beyond the planet's atmosphere in outer space;
- they can actively influence their evolution by overcoming natural selection, altering the culture and environment in which they live, and manipulating their genomes;
- they are aware of themselves as connected to one another and to all life, and as stewards of the Planet Earth.

Classify this new species as *Homo electric*. Some may object to calling it a new species, arguing that it has not undergone the genetic and biological transformation to make it so. Perhaps the evidence of such a transformation is not apparent yet. But in any event, this traditional physical view of human evolution overlooks the essence of what makes us humans such unique animals. We are more than just flesh and bones and organs and genes. We are cultural beings in an electrified world. We are who we think we are and who we aspire to be.

Planetary awareness -- our conscious connection with all humans, with all living things -- is a defining trait of *electrics.* It is an evolutionary inevitability that the world of these new humans will become as one. But many s*apiens* will not readily relinquish their territoriality, patriarchy, and exploitation, nor cease attempting to exert divisive and controlling powers over human society. That is their very nature as *sapiens.* It is the source of much of the social disruption we are experiencing today. Eventually it could lead to interspecies competition and conflict, in the halls of government and in the streets. But neither ballots nor bullets can stop evolution.

Ultimately, a unified Planet Electric will emerge. I will not see it in my lifetime. You may not either. But evolution is patient, and relentless. It has left behind *Homo erectus, Homo neanderthalensis* and all other previous human species, as it surely will leave behind *Homo sapiens*.

Human evolution has not come to an end. For perhaps a million years, fire was the foundation of human existence. But in less than two centuries, we have transitioned to electricity as that foundation. We are living in a time of epochal change. A new human species is emerging: *Homo electric.*

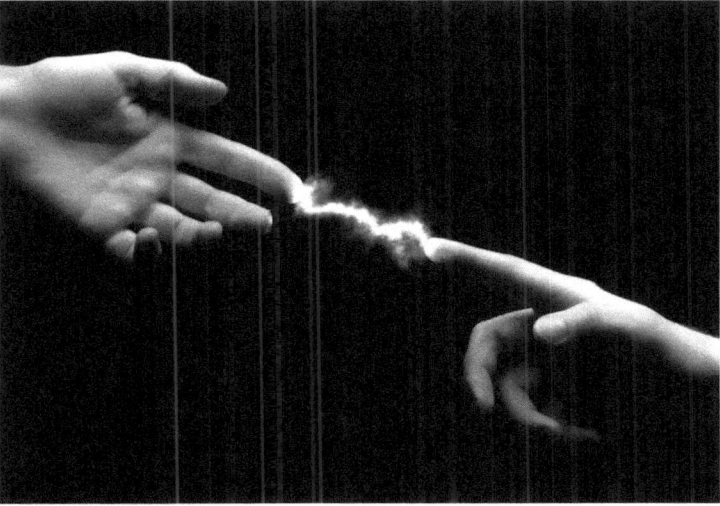

I am *electric*. Are you?

AFTERWORD

It's been over five years since my lightbulb moment. Nothing I've read or seen or heard in my journey since then has contradicted my insight: just as controlling fire was a pivotal development in human history and enabled the evolution of archaic humans, so too is controlling electricity a pivotal development empowering the ongoing evolution of humans today.

Having written this book, I am convinced more than ever of the validity of that insight. It has revealed to me who I am as a human being in an electrified world. The first thing I see in the morning are the digits displayed on my electric clock; the last thing at night is my bedside lamp, before I turn it off and go to sleep. A massive electric infrastructure maintains this world 24/7. When I wake up the next day, faces will still appear on my screen to tell me what's going on, the food in my refrigerator will still be cold, and my money will still be entered in the servers which hold my bank accounts.

Some places still lack electricity. A project is underway to electrify half of the 600 million Africans without it. A woman in Tanzania, Mwajuma Mohamed, said of a previous effort: "When we got electricity, it was like we were normal people suddenly" (*The New York Times*, January 27, 2025).

The evolutionary importance of fire is unquestioned. As Yale University paleoanthropologist Jessica Thompson said in a TEDx talk in 2022, fire enabled "the transformation of our species from one that was more embedded within its ecosystem to one that increasingly took control over its own evolution and began to engineer the world around them." Electricity has exponentially accelerated that process.

She went on to say that "arguably fire has been the most transformative of all of our tools." True enough in the ancient world, but not nearly as transformative as electricity has been in the modern world. After they learned to control fire, early humans continued to live as hunter-gatherers for hundreds of thousands of years. But since we began using electricity less than two hundred years ago, it has displaced fire as our primary source of energy and provided us with a myriad of tools and uses that would fill this book if I tried to list them.

Donald Johanson is the founding director of the Institute of Human Origins at Arizona State University. He has a long view of human history, as the man who discovered the fossil "Lucy." Speaking at an IHO conference in New York in 2023, Don said: "We are involved and faced at the moment with I think a very critical moment in terms of human evolution."

He continued: "It is time for us as a species to become even more introspective about who we are, where we're going and how we can preserve this incredible experience of being on this pale blue dot," as his good friend Carl Sagan called our planet. Will *Homo sapiens* try to solve the climate crisis, or exacerbate it? Or will a new species *Homo electric* take the evolutionary baton in the race for survival? A species with the planetary awareness expressed by Don?

In this book I have tried to understand and explain how we "are being evolved." I hope that anthropologists, geneticists, biologists, journalists, artists, poets and others will join me in exploring the vast terrain of our electrified world and better understanding the people who are emerging to inhabit it.

And I hope this book has switched on a lightbulb for you.

ABOUT THE AUTHOR

Steven Reed Nelson is not an academic or scientific expert on evolution, but an "outsider." As was the employee of the Swiss patent office who deduced that $E = mc^2$. As was the Czech monk who discovered the principles of heredity.

Steve does not claim to be an Einstein or a Mendel, but he is a rigorous thinker: a math major at Cornell University and a prize-winning student at Harvard Law School. He was chosen to do anthropological research in the Peruvian Andes, where he experienced a world without electricity.

Since then he has lived a life electric and eclectic, often ahead of the curve and anticipating trends. He has been an entrepreneur and marketer of cutting-edge applications of electricity such as broadband internet, solar energy, computer software, rock concerts and video production. Immersed in emerging electric technologies, at the same time he observed their effects on humans with the eye of an anthropologist.

Steve lives in the hills of western Massachusetts, where he sees the sun going down and the world spinning 'round.

Jan and Steve, self-portrait with video feedback loop, 1970

ACKNOWLEDGMENTS

My wife Jan Lewis Nelson was nearing the end of a thirty-year battle with breast cancer when I told her my idea about electricity and evolution. She encouraged me to write a book. I promised her I would. After she passed away in 2020, I did write an earlier version of this book, but my thoughts often were elsewhere, and I was not satisfied with it. I returned to the project in 2023. Had she been able to read the manuscript, no doubt Jan, an author herself, would have found words and passages in need of editing. Readers of the book and I would have benefited from her input, as I benefited in countless ways during the fifty years she and I were together.

Several people did read all or part of the manuscript, and likewise encouraged me. I am thankful for their support. If my brother Dr. Peter L. Nelson did not find credible my premise about the evolutionary parallel between fire and electricity, I might not have continued. Thanks to my son Nate, who is always there for me, and to readers Steve Haken, Barbara Newman, Marty Perlmutter, Michael Pitkow and Miki Raver.

Originally I intended to self-publish it, but along the way I got distracted by the bright shiny object of traditional publishing. My inquiries for representation by a literary agent were rejected or ignored multiple times until Claire Genus signed me. Some of the editors to whom she submitted our book proposal did find it "interesting" or even "fascinating," but questioned: who was I to write it? I'm grateful to Claire for expressing confidence in me and encouraging me to self-publish it after all. Now that you have read it, I'll leave it up to you to decide if I was qualified to write this book.

BIBLIOGRAPHY

Bloom, Howard, *Global Brain* (John Wiley & Sons, 2000)

Darwin, Charles, *On the Origin of Species* (1859)

-------, *The Descent of Man, and Selection in Relation to Sex* (1871)

-------, *The Voyage of the Beagle* (1839)

Harari, Juval Noah, *Sapiens: A Brief History* of *Humankind* (Harper Perennial, 2018)

Henrich, Joseph, *The Secret of Our Success: How Culture Is Driving Human Evolution, Domesticating Our Species, And Making Us Smarter* (Princeton University Press, 2016)

Jonnes, Jill, *Empires of Light: Edison, Tesla, Westinghouse, and the Race to Electrify The World* (Random House Trade Paperbacks, 2003)

Moore, David S., *The Developing Genome: An Introduction to Behavioral Epigenetics* (Oxford University Press, 2015)

Nelson, Peter L., socsci.biz/paper1.pdf, socsci.biz/paper2.pdf

Port, Ian S., *The Birth of Loud: Leo Fender, Les Paul, and the Guitar-Pioneering Rivalry That Shaped Rock 'n' Roll* (Scribner, 2019)

Smith, Adam, *The Wealth of Nations* (1776)

Wilson, David Sloan, *Evolution for Everyone* (Bantam Dell, 2007)

-------, *This View of Life: Completing the Darwinian Revolution* (Pantheon, 2019)

Wrangham, Richard, *Catching Fire: How Cooking Made Us Human* (Basic Books, 2010)

Yandell, Kay, *Telegraphies: Indigeneity, Identity and Nation in America's Nineteenth-Century Virtual Realm* (Oxford University Press, 2019)

Yassini-Fard, Rouzbeh, with Stewart Schley, *The Accidental Network: How a Small Company Sparked a Global Broadband Transformation* (West Virginia University Press, 2025)

CREDITS

All images in this book are either by the author, in the public domain, used with permission or licensed.

53: Aidonis *et al.*, doi. 10.1016/j.jasrep.2023.104206

54: Cicero Moraes *et alii*, Creative Commons 4.0

55: hairymuseummatt (original photo), DrMikeBaxter (derivative work), Creative Commons 2.0

56: (top and bottom) Neanderthal-Museum, Mettmann, Germany, Creative Commons 4.0

58: Neanderthal-Museum, Mettmann, Germany, Creative Commons 4.0

63: DJ, Creative Commons 2.0

64. 123rf.com

66: OER Commons

67: pexels.com

69: Charles Levy (1945), U.S. National Archives and Records Admin.

70: U.S. Rural Electrification Administration

72: (top) U.S. National Park Service
(bottom) Francis Godolphin Osbourne Stuart

77: Alexander Gardner (1865)

79: Detail from Benjamin West painting, *Benjamin Franklin Drawing Electricity From the Sky* (1816)

80: (top) public domain, photographer unknown (c. 1850s)
(bottom) public domain

88: Kaamran Hafeez via CartoonStock

90: Stadtpflaenzchen, Creative Commons 1.0

92: *The Hornet* magazine, March 22, 1871

94: Frontispiece, *Mendel's Principles of Heredity: A Defence* (1902)

97: Medical image ineligible for copyright

108: Wasin Pummarin © 123rf.com

112: Daniil Peshkov © 123rf.com

114: © <a href='https://www.123rf.com/profile_virtosmedia'>-virtosmedia</a>, <a href='https://www.123rf.com/free-images/'>-123RF Free Images</a>

116: Liu Zishan © 123rf.com

122: Photo from the Northeast Solar Energy Center

124: Didier, Petit et Cie (1859), Creative Commons 1.0

126: U.S. SAGE Air Defense System

130: Photo by the author

135: Photo by the author

140: Asa Wilson, Creative Commons 2.0

146: Steven Bennett, Castellar Zoo, Creative Commons 2.0

148: U.S. government photos

154: Publicity photo

157: National Museum of Slovenia, Creative Commons Zero

158: Photo by the author

161: MGM / Verve Records publicity photo

163: Poster designed by the author

164: (top) Photo by the author
(bottom) Publicity photo

166: Eadward Muybridge

170: The Metropolitan Museum of Art

180: 123rf.com

184: Pop Nukoonrat © 123rf.com

186: Photo by the author

190: Bill Anders, National Aeronautics and Space Administration

199: Maksym Zalecskyy © 123rf.com

202: Photo by Jan Lewis Nelson

204: Photo by the author

206: Jeff Stahler / Andrews McMeel Syndication

217: Buzz Aldrin, National Aeronautics and Space Administration

The last line of "About the Author" is adapted from the lyrics to "The Fool on the Hill" by The Beatles.

Cover concept and interior design by the author.

INDEX

Footprint on the moon

www.ingramcontent.com/pod-product-compliance
Lightning Source LLC
Chambersburg PA
CBHW041256040426
42334CB00028BA/3033